M337 Unit A2
Mathematics: A Third Level Course

COMPLEX ANALYSIS
UNIT A2 COMPLEX FUNCTIONS

Prepared by the Course Team

> Before working through this text, make sure that you have read the
> *Course Guide* for M337 Complex Analysis.

The Open University, Walton Hall, Milton Keynes, MK7 6AA.

First published 1993. Reprinted 1995, 1998, 2001, 2008

Copyright © 1993 The Open University

All rights reserved. No part of this publication may be reproduced, stored in a retrieval system or transmitted in any form or by any means, without written permission from the publisher or a licence from the Copyright Licensing Agency Limited. Details of such licences (for reprographic reproduction) may be obtained from the Copyright Licensing Agency Ltd of 90 Tottenham Court Road, London, W1P 9HE.

Edited, designed and typeset by the Open University using the Open University T$_E$X System.

Printed in Malta by Gutenberg Press Limited.

ISBN 0 7492 2176 3

This text forms part of an Open University Third Level Course. If you would like a copy of *Studying with The Open University*, please write to the Central Enquiry Service, PO Box 200, The Open University, Walton Hall, Milton Keynes, MK7 6YZ. If you have not already enrolled on the Course and would like to buy this or other Open University material, please write to Open University Educational Enterprises Ltd, 12 Cofferidge Close, Stony Stratford, Milton Keynes, MK11 1BY, United Kingdom.

CONTENTS

Introduction		4
	Study guide	4
1	What is a Complex Function?	5
	1.1 Defining functions	5
	1.2 The image of a function	6
	1.3 Sums, products and quotients of functions	8
	1.4 Composite functions	9
	1.5 Inverse functions	10
2	Special Types of Complex Function	14
	2.1 Functions with codomain $\mathbb{R}$	14
	2.2 Functions with domain a subset of $\mathbb{R}$	16
3	Images of Grids (audio-tape)	21
4	Exponential, Trigonometric and Hyperbolic Functions	26
	4.1 The exponential function	26
	4.2 Trigonometric functions	32
	4.3 Hyperbolic functions	36
5	Logarithms and Powers	38
	5.1 Logarithms of complex numbers	38
	5.2 Powers of complex numbers	43
Exercises		44
Solutions to the Problems		47
Solutions to the Exercises		54

INTRODUCTION

In *Unit A1* we introduced complex numbers and described a number of fundamental operations with them; in particular, we investigated the solution of equations involving complex numbers. We now study *complex functions* and we find that many of the standard real functions (polynomial functions, rational functions, the exponential, trigonometric and hyperbolic functions) have complex analogues with remarkable geometric properties.

In Section 1 we establish the basic language associated with complex functions, such as sums, products, quotients and composites of functions. We also discuss the problems associated with forming inverses of complex functions.

In Section 2 we discuss two special types of complex function, namely those with codomain $\mathbb{R}$ and those with domain a subset of $\mathbb{R}$. Each of these types of function has a role to play in our understanding of the geometric nature of a complex function, which we study in Section 3. There we examine some particular complex functions in detail and sketch the images of various 'grids' under these functions.

In Section 4 we introduce the complex exponential function and describe its geometric properties. The complex trigonometric functions and hyperbolic functions are then defined in terms of this exponential function.

Finally, in Section 5, we introduce the complex logarithm function, which will have an important part to play in complex integration; we also discuss complex powers.

Study guide

You should find that Section 1 is mainly revision and you should aim to work through it fairly quickly. Section 2 and Section 3 (the audio-tape section) are closely related and not too long, so you might try to study them in one session. Sections 4 and 5 are also closely related, and they contain a good deal of basic technical material which will be used throughout the course.

Associated with this unit is a segment of the Video Tape for the course. Although this unit text is self-contained, access to the video tape will enhance your understanding. Suitable points at which to view the video tape are indicated by a symbol placed in the margin.

1 WHAT IS A COMPLEX FUNCTION?

After working through this section, you should be able to:

(a) determine the domain and rule of the *sum, product* and *quotient* of two complex functions;

(b) determine the domain and rule of the *composite* of two complex functions;

(c) determine whether a given complex function has an *inverse* function, and find that inverse function in suitable cases.

1.1 Defining functions

The main aim of complex analysis is to extend the theory of calculus to include functions of a complex variable; such functions are called *complex functions*.

Definitions A **complex function** f is defined by specifying

(a) two sets A and B in the complex plane $\mathbb{C}$, and

(b) a rule which associates with each number z in A a unique number w in B; we write $w = f(z)$.

The set A is called the **domain** of the function f and the set B is called the **codomain** of f. The number w is called the **image of z under f**, or the **value of f at z**.

Notice that a *real function* is a particular type of complex function, in which A and B are both subsets of $\mathbb{R}$ (since $\mathbb{R} \subseteq \mathbb{C}$).

For example, the expression

$$f : \mathbb{C} \longrightarrow \mathbb{C}$$
$$z \longmapsto z^2 \qquad (1.1)$$

defines a complex function with $A = \mathbb{C}$ and $B = \mathbb{C}$, and with rule given by $f(z) = z^2$. Under this function f, the image of $z = 2i$ is $w = f(2i) = (2i)^2 = -4$, and, similarly, the image of $1 - i$ is $-2i$. These values are plotted in Figure 1.1; it shows the points $2i$ and $1 - i$ in the domain (the z-plane) and the corresponding images in the codomain (the w-plane).

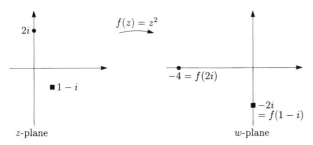

Figure 1.1

We shall often use this type of diagrammatic representation for specific complex functions, whereas a general complex function $f : A \longrightarrow B$ may be represented by a diagram in which the sets A and B appear as 'blobs' (Figure 1.2).

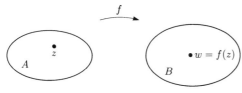

Figure 1.2

These diagrams make it clear why some texts refer to the codomain B as the 'range' or 'target'.

Usually we shall refer to a complex function f simply as a 'function', unless there is some particular reason to emphasize that f is a complex function. Other commonly used words for function are 'map', 'mapping' and 'transformation'.

The notation (1.1) can be written more concisely in any one of several forms

$$f(z) = z^2 \quad (z \in \mathbb{C}), \qquad z \longmapsto z^2 \quad (z \in \mathbb{C}),$$

where it is assumed that the codomain is $\mathbb{C}$, or

$$f(z) = z^2, \qquad z \longmapsto z^2,$$

where it is assumed that both the domain and codomain are $\mathbb{C}$. To avoid uncertainty, we adopt the following convention.

Convention

When a function f is specified *just* by its rule, it is to be understood that the domain of f is the set of all complex numbers to which the rule is applicable, and the codomain of f is $\mathbb{C}$.

For example, the function

$$f(z) = \frac{1}{z - i}$$

has domain $\mathbb{C} - \{i\}$ because $1/(z - i)$ is defined for all complex numbers z, other than $z = i$. Its codomain is $\mathbb{C}$.

Problem 1.1

With the above convention, write down the domain and codomain of each of the following functions.

(a) $f(z) = z + 2$ (b) $f(z) = \dfrac{z}{z + 2}$ (c) $f(z) = \operatorname{Arg} z$

(d) $f(z) = \dfrac{1}{z^2 + 1}$

1.2 The image of a function

If a function f has domain A and codomain B, then for each z in A the image $w = f(z)$ is in B. However, it is not necessarily true that, for each w in B, there is some z in A such that $f(z) = w$. For example, if $f(z) = 1/(z - i)$, then the domain of f is $\mathbb{C} - \{i\}$ and the codomain is $\mathbb{C}$ (by the convention). However, there is no point z in the domain of f which maps to the point 0 (in the codomain); that is, there is no z in $\mathbb{C} - \{i\}$ such that $f(z) = 0$.

Definition Given a function $f : A \longrightarrow B$, the **image of f**, written $f(A)$, is the set of all values $f(z)$, where $z \in A$ (Figure 1.3). Thus

$$f(A) = \{f(z) : z \in A\}.$$

If $f(A) = B$, then the function f is said to be **onto**.

Some texts use 'surjective' rather than 'onto'.

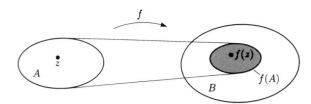

Figure 1.3 $f(A)$, the image of f

The function depicted in Figure 1.3 is *not* onto because $f(A) \neq B$.

Example 1.1

Determine the image of the function $f(z) = 1/(z-i)$.

Solution

The domain of f is $A = \mathbb{C} - \{i\}$ and $f(z) = \dfrac{1}{z-i}$, so we have

$$
\begin{aligned}
f(A) &= \left\{ \frac{1}{z-i} : z \in \mathbb{C} - \{i\} \right\} \\
&= \left\{ w = \frac{1}{z-i} : z \neq i \right\} \\
&= \left\{ w : z = i + \frac{1}{w} \neq i \right\} \quad \left(w = \frac{1}{z-i} \iff z = i + \frac{1}{w} \right) \\
&= \{ w : w \neq 0 \} \quad (1/w \text{ exists and is non-zero} \iff w \neq 0) \\
&= \mathbb{C} - \{0\}. \quad \blacksquare
\end{aligned}
$$

The aim is to express $f(A)$ in the form $\{w : \text{condition on } w\}$ and hence identify the set.

Remark In this case perhaps you could 'see' what the image of f is; however, you need to be able to write down the argument.

Problem 1.2

For each of the following functions f, determine the image of f.

(a) $f(z) = 3iz$ (b) $f(z) = \dfrac{3z+1}{z+i}$ (c) $f(z) = \operatorname{Im} z$

Several important functions, such as

$$z \longmapsto \operatorname{Re} z, \quad z \longmapsto \operatorname{Im} z, \quad z \longmapsto |z| \quad \text{and} \quad z \longmapsto \operatorname{Arg} z$$

have images which are subsets of the real line. For example, the function $f(z) = \operatorname{Re} z$ has image $f(\mathbb{C}) = \mathbb{R}$, the whole real line (Figure 1.4). These are often referred to as **real-valued functions** (of a complex variable).

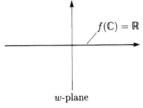

Figure 1.4 The image of $f(z) = \operatorname{Re} z$

Problem 1.3

Write down the image of each of the following functions.

(a) $f(z) = |z|$ (b) $f(z) = \operatorname{Arg} z$

1.3 Sums, products and quotients of functions

Let f and g be the functions

$$f(z) = \frac{1}{z} \quad (z \in \mathbb{C} - \{0\}) \quad \text{and} \quad g(z) = z^2 + 1 \quad (z \in \mathbb{C}).$$

The expressions $f + g$, fg and f/g are used to denote the following functions

$$(f + g)(z) = f(z) + g(z) = \frac{1}{z} + (z^2 + 1) \quad (z \in \mathbb{C} - \{0\});$$

$$(fg)(z) = f(z)g(z) = \frac{1}{z}(z^2 + 1) \quad (z \in \mathbb{C} - \{0\});$$

$$(f/g)(z) = f(z)/g(z) = \frac{1}{z}\left(\frac{1}{z^2 + 1}\right) \quad (z \in \mathbb{C} - \{0, i, -i\}).$$

The domains of $f + g$ and fg include only those points at which both f and g are defined. When forming the quotient f/g, we must also exclude from the domain those points z at which $g(z) = 0$. Such points z are called **zeros** of g.

Definitions Let $f : A \longrightarrow \mathbb{C}$ and $g : B \longrightarrow \mathbb{C}$ be functions. Then the **sum** $f + g$ is the function with domain $A \cap B$ and rule

$$(f + g)(z) = f(z) + g(z);$$

the **multiple** λf, where $\lambda \in \mathbb{C}$, is the function with domain A and rule

$$(\lambda f)(z) = \lambda f(z);$$

the **product** fg is the function with domain $A \cap B$ and rule

$$(fg)(z) = f(z)g(z);$$

the **quotient** f/g is the function with domain $A \cap B - \{z : g(z) = 0\}$ and rule

$$(f/g)(z) = f(z)/g(z).$$

Remark If $f(z) = z$ and $g(z) = 1/z$, then fg has domain $\mathbb{C} \cap (\mathbb{C} - \{0\}) = \mathbb{C} - \{0\}$, and rule $(fg)(z) = 1$. Thus the domain of fg is *not* the largest set on which the rule of fg is applicable. In this case the domain of fg is given by the definition of the product fg, so the convention on page 6 does not apply.

Problem 1.4

Let f and g be the functions

$$f(z) = \frac{1}{z} \quad (z \in \mathbb{C} - \{0\}) \quad \text{and} \quad g(z) = \frac{z + 3i}{z^2 - z} \quad (z \in \mathbb{C} - \{0, 1\}).$$

Determine the domain and rule of each of the following functions.
(a) $f + g$ (b) fg (c) f/g

Starting from the two basic functions $z \longmapsto 1$ and $z \longmapsto z$, we can build up any **polynomial function** of degree n

$$p(z) = a_0 + a_1 z + a_2 z^2 + \cdots + a_n z^n,$$

where $a_0, a_1, \ldots, a_n \in \mathbb{C}, a_n \neq 0$, by forming suitable sums, multiples and products. The domain of any polynomial function is $\mathbb{C}$. Allowing quotients also, we can build up any **rational function**; that is, any function of the form

$$f(z) = \frac{p(z)}{q(z)},$$

where p and q are polynomial functions. It follows from the definition of 'quotient' that the domain of such a rational function is $\mathbb{C} - \{z : q(z) = 0\}$.

For example, the rational function
$$f(z) = \frac{z}{z^2 + 1}$$
has domain $\mathbb{C} - \{i, -i\}$.

1.4 Composite functions

Let f and g be functions. The composite function $g \circ f$ is obtained by applying first f and then g. Thus the rule of $g \circ f$ is

$$(g \circ f)(z) = g(f(z)).$$

The process of forming $g \circ f$ is called 'composition of functions' or 'composing functions'.

For example, if

$$f(z) = \frac{1}{z} \quad (z \in \mathbb{C} - \{0\}) \quad \text{and} \quad g(z) = z^2 + 1 \quad (z \in \mathbb{C}),$$

then the rule of $g \circ f$ is

$$(g \circ f)(z) = g(1/z) = (1/z)^2 + 1,$$

whereas the rule of $f \circ g$ is

$$(f \circ g)(z) = f(z^2 + 1) = 1/(z^2 + 1).$$

But what is the domain of a composite function? In general, if $f: A \longrightarrow \mathbb{C}$ and $g: B \longrightarrow \mathbb{C}$, then the value $g(f(z))$ is defined if and only if

z lies in A and $f(z)$ lies in B.

Thus, if z_1 and z_2 are elements of A such that $f(z_1) \in B$ but $f(z_2) \notin B$, then $g(f(z))$ is defined for $z = z_1$ but not for $z = z_2$, as indicated in Figure 1.5.

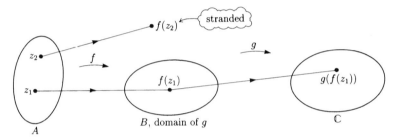

Figure 1.5

We define the domain of $g \circ f$ to be consistent with this.

Definition Let $f: A \longrightarrow \mathbb{C}$ and $g: B \longrightarrow \mathbb{C}$ be complex functions. Then the **composite function** $g \circ f$ has domain

$$\{z \in A : f(z) \in B\}$$

and rule

$$(g \circ f)(z) = g(f(z)).$$

To find the domain of $g \circ f$, we remove from A (the domain of f) each point whose image under f is not in B (the domain of g). For example, if

$$f(z) = z^2 + i \quad (z \in \mathbb{C}) \quad \text{and} \quad g(z) = \frac{1}{z - i} \quad (z \in \mathbb{C} - \{i\}),$$

Thus, the domain of $g \circ f$ may be written as

$$A - \{z : f(z) \notin B\}.$$

then the domain of $g \circ f$ is

$$\mathbb{C} - \{z \in \mathbb{C} : z^2 + i \notin \mathbb{C} - \{i\}\} = \mathbb{C} - \{z \in \mathbb{C} : z^2 + i = i\} = \mathbb{C} - \{0\}.$$

In practice, it often happens that the image of f is contained in B — that is, $f(A) \subseteq B$ — and in this case the domain of $g \circ f$ is A itself. (This always happens when the domain of g is $\mathbb{C}$, of course.) For example, if

Some texts *require* that $f(A) \subseteq B$ in the definition of $g \circ f$.

$$f(z) = \frac{1}{z - i} \quad (z \in \mathbb{C} - \{i\}) \quad \text{and} \quad g(z) = \text{Arg } z \quad (z \in \mathbb{C} - \{0\}),$$

then $A = \mathbb{C} - \{i\}$, $B = \mathbb{C} - \{0\}$; also, as you saw in Example 1.1,

$$f(\mathbb{C} - \{i\}) = \mathbb{C} - \{0\},$$

which is (contained in) the domain of g. Thus the domain of $g \circ f$ is $\mathbb{C} - \{i\}$.

$g(f(z)) = \text{Arg}(1/(z - i))$.

Problem 1.5

Let f and g be the functions

$$f(z) = \frac{1}{z} \quad (z \in \mathbb{C} - \{0\}) \quad \text{and} \quad g(z) = \frac{z + 3i}{z^2 - z} \quad (z \in \mathbb{C} - \{0, 1\}).$$

Determine the domain and rule of each of the following functions.

(a) $g \circ f$ (b) $f \circ g$

1.5 Inverse functions

Let f be the function

$$f(z) = 3z \quad (z \in \mathbb{C}).$$

Then, for each number w in $\mathbb{C}$, there is a unique number $z = \frac{1}{3}w$ in the domain of f such that

$$f(z) = f(\tfrac{1}{3}w) = 3 \times \tfrac{1}{3}w = w.$$

The corresponding function $g : w \longmapsto \frac{1}{3}w$ is called the *inverse function* of f because it undoes the effect of f. To be precise

$$g(f(z)) = \tfrac{1}{3}f(z) = \tfrac{1}{3} \times 3z = z, \quad z \in \mathbb{C};$$

similarly, f undoes g

$$f(g(w)) = 3g(w) = 3 \times \tfrac{1}{3}w = w, \quad w \in \mathbb{C}.$$

The inverse function of f is denoted by f^{-1}. The functions f and f^{-1} are illustrated in Figure 1.6

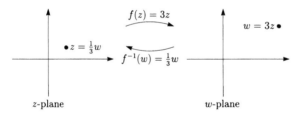

Figure 1.6

Not every function has such an inverse function. For example, consider the function

$$f(z) = z^2 \quad (z \in \mathbb{C}).$$

Since $f(2) = 4$ and $f(-2) = 4$, we cannot assign a unique value z in the domain of f such that $f(z) = 4$ (Figure 1.7).

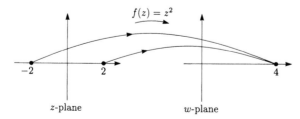

Figure 1.7

The problem here is that the function f is not *one-one*. In general, it is possible to define the inverse of a function only if that function is one-one.

Definition The function $f : A \longrightarrow B$ is **one-one** if the images under f of distinct points in A are also distinct; that is,

if $z_1, z_2 \in A$ and $z_1 \neq z_2$, then $f(z_1) \neq f(z_2)$.

Some texts use 'injective' rather than 'one-one'. Also, note that a function which is not one-one is said to be **many-one**.

An equivalent statement of this condition is that

if $w \in f(A)$, then there is a *unique* z in A such that $f(z) = w$

(Figure 1.8).

This uniqueness makes it possible to define the inverse function, f^{-1}, of f with domain $f(A)$ (Figure 1.9).

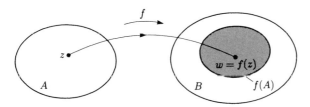

Figure 1.8

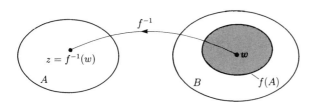

Figure 1.9

Definition Let $f : A \longrightarrow B$ be a one-one function. Then the **inverse function**, f^{-1}, of f has domain $f(A)$ and rule

$f^{-1}(w) = z,$ where $w = f(z)$.

Thus, there are two ways of proving that a function has an inverse function.

Strategy for proving that a function has an inverse function

To prove that a function f has an inverse function

EITHER

prove that f is one-one directly by showing that

if $z_1 \neq z_2$, then $f(z_1) \neq f(z_2)$

OR

determine the image $f(A)$ and show that for each $w \in f(A)$ there is a unique $z \in A$ such that $f(z) = w$.

This statement is equivalent to
$f(z_1) = f(z_2) \implies z_1 = z_2$.

One way of demonstrating that 'there is a unique $z \in A$ such that $f(z) = w$' is to *find* the rule for f^{-1} (if this is possible). For some functions this can be done by solving the equation $w = f(z)$ to obtain a unique z in terms of w, as illustrated in the next example.

We adopt the second strategy with this approach whenever it is possible, for it has the advantage that we thereby specify the function f^{-1}.

Example 1.2

Prove that the function

$$f(z) = \frac{1}{z-i} \quad (z \in \mathbb{C} - \{i\})$$

has an inverse function, and determine the domain and rule of f^{-1}.

Solution

First we determine the image of f. This is

$$f(\mathbb{C} - \{i\}) = \mathbb{C} - \{0\} \quad \text{(from Example 1.1)}.$$

Now, for each $w \in \mathbb{C} - \{0\}$, we wish to solve the equation

$$w = \frac{1}{z-i}$$

to obtain a *unique* solution z in $\mathbb{C} - \{i\}$. This is achieved by the rearrangement

$$z = i + \frac{1}{w}.$$

Thus f is a one-one function with image $\mathbb{C} - \{0\}$. Hence f has an inverse function f^{-1} with domain $\mathbb{C} - \{0\}$ and rule

$$f^{-1}(w) = i + \frac{1}{w} \quad (w \in \mathbb{C} - \{0\}). \quad \blacksquare$$

Remarks

1 Notice that in Example 1.2, the domain of the inverse function f^{-1} is equal to the largest set for which the rule of f^{-1} is applicable. However, this does not always occur. For example, the inverse function of the function

$$f(z) = \frac{1}{z-i} \quad (|z| < 1)$$

has rule $f^{-1}(w) = i + 1/w$, but its domain is a proper subset of $\mathbb{C} - \{0\}$. (In fact, the domain of f^{-1} is $\{w : \operatorname{Im} w > \frac{1}{2}\}$.)

2 Usually when defining a function, we write z for the domain variable. To conform to this practice, we could rewrite the inverse function f^{-1} in Example 1.2 in the form

$$f^{-1}(z) = i + \frac{1}{z} \quad (z \in \mathbb{C} - \{0\}),$$

after the required algebraic manipulations have been completed. It is *not* necessary to do this as a matter of routine.

Problem 1.6

Prove that the function

$$f(z) = \frac{3z+1}{z+i} \quad (z \in \mathbb{C} - \{i\})$$

has an inverse function f^{-1}, and determine the domain and rule of f^{-1}. (See Problem 1.2(b).)

As we pointed out above, the function $f(z) = z^2$ is not one-one (on $\mathbb{C}$), and so it does not have an inverse function. One way round this difficulty is to reduce the domain of f (without changing the rule) so as to make the resulting function one-one. Note that reducing the domain of a function leads to a *new* function — a **restriction** of the original function — which should really be denoted by a different letter. In practice, we usually retain the same letter, particularly if the original domain is no longer under discussion. Here is an example of this.

Example 1.3

Let $A = \{0\} \cup \{z : -\tfrac{1}{2}\pi < \operatorname{Arg} z \leq \tfrac{1}{2}\pi\}$, as shown in Figure 1.10.
Prove that the function
$$f(z) = z^2 \qquad (z \in A)$$
has an inverse function f^{-1}, and determine the domain and rule of f^{-1}.

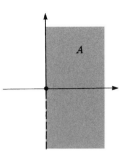

Figure 1.10

Solution

First we determine the image of f:
$$\begin{aligned}
f(A) &= \{z^2 : z \in A\} \\
&= \{0\} \cup \{w = z^2 : -\tfrac{1}{2}\pi < \operatorname{Arg} z \leq \tfrac{1}{2}\pi\} \quad (0^2 = 0) \\
&= \{0\} \cup \{w = r^2(\cos 2\theta + i \sin 2\theta) : r > 0, -\tfrac{1}{2}\pi < \theta \leq \tfrac{1}{2}\pi\} \\
&\qquad (z = r(\cos\theta + i \sin\theta)) \\
&= \{0\} \cup \{w = \rho(\cos\phi + i \sin\phi) : \rho > 0, -\pi < \phi \leq \pi\} \\
&\qquad (\rho = r^2, \phi = 2\theta) \\
&= \mathbb{C}.
\end{aligned}$$

Now, for each $w \in \mathbb{C}$, we wish to solve the equation
$$w = z^2 \tag{1.2}$$
to obtain a *unique* solution z in A. If $w = 0$, then Equation (1.2) has the unique solution $z = 0$. On the other hand, if $w \neq 0$, then w can be written in the form
$$w = \rho(\cos\phi + i \sin\phi), \quad \text{where } \rho > 0 \text{ and } -\pi < \phi \leq \pi,$$
and Equation (1.2) has exactly two solutions:
$$z_0 = \rho^{1/2}\left(\cos\tfrac{1}{2}\phi + i \sin\tfrac{1}{2}\phi\right)$$
and
$$z_1 = \rho^{1/2}\left(\cos\left(\tfrac{1}{2}\phi + \pi\right) + i \sin\left(\tfrac{1}{2}\phi + \pi\right)\right),$$
by Theorem 3.1 of *Unit A1* (Figure 1.11). Clearly $z_0 \in A$, since $-\tfrac{1}{2}\pi < \tfrac{1}{2}\phi \leq \tfrac{1}{2}\pi$, whereas $z_1 \notin A$.

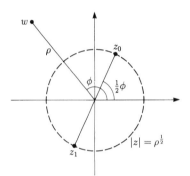

Figure 1.11

Thus f is a one-one function with image $\mathbb{C}$. Hence f has an inverse function f^{-1} with domain $\mathbb{C}$ and rule
$$f^{-1}(w) = \begin{cases} \rho^{1/2}\left(\cos\tfrac{1}{2}\phi + i \sin\tfrac{1}{2}\phi\right), & w = \rho(\cos\phi + i \sin\phi), \rho > 0, -\pi < \phi \leq \pi, \\ 0, & w = 0. \end{cases} \blacksquare$$

Remark In this solution we chose ϕ to satisfy $-\pi < \phi \leq \pi$ so that $-\tfrac{1}{2}\pi < \tfrac{1}{2}\phi \leq \tfrac{1}{2}\pi$, and hence $z_0 \in A$. Since $\phi = \operatorname{Arg} w$, for $w \neq 0$, it follows that z_0 is the principal square root, $\sqrt{w}$, of w. Since we defined $\sqrt{0} = 0$, the rule for f^{-1} can be written in the form
$$f^{-1}(w) = \sqrt{w} \qquad (w \in \mathbb{C}).$$

See *Unit A1*, Section 3.

The set A in Example 1.3 is not the only one on which the function $z \longmapsto z^2$ is one-one with image $\mathbb{C}$. In the following problem you are asked to investigate another such set.

Problem 1.7

Let $A = \{0\} \cup \{z : 0 \leq \operatorname{Arg} z < \pi\}$. Prove that the function
$$f(z) = z^2 \qquad (z \in A)$$
has an inverse function f^{-1}, and determine the domain and rule of f^{-1}.

2 SPECIAL TYPES OF COMPLEX FUNCTION

After working through this section, you should be able to:

(a) find the *real* and *imaginary parts* of a complex function and sketch their graphs in simple cases;

(b) sketch a *path*;

(c) obtain (where possible) the equation of a path by eliminating the parameter from its parametrization;

(d) find the *image* under a function of a path in simple cases;

(e) use the table of standard parametrizations.

2.1 Functions with codomain $\mathbb{R}$

In Section 1 we pointed out that various common functions, such as $z \longmapsto |z|$, have their images in $\mathbb{R}$ and are therefore called **real-valued functions**. Because the image of such a function is in $\mathbb{R}$ we do not have to resort to the z-plane/w-plane representation of the function. In fact, we can sketch the graph of a real-valued function by introducing a third axis (here called the s-axis), which is at right angles to the complex plane (z-plane). Figure 2.1 shows the graph of the function $z \longmapsto |z|$; it is the surface $s = |z| = \sqrt{x^2 + y^2}$.

We have avoided calling the new axis the z-axis because of possible confusion with the complex variable $z = x + iy$.

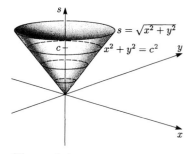

Figure 2.1

We have clarified the nature of this surface by indicating various curves at constant height. These are the curves given by the intersection of the surface $s = \sqrt{x^2 + y^2}$ with horizontal planes of the form $s = c$, where c is constant; they are evidently circles of the form $x^2 + y^2 = c^2$, at height c. This explains the conical shape of the surface.

Figure 2.2(a) shows the spiral-like surface $s = \text{Arg } z$. It is obtained by lifting (that is, translating vertically) each of the rays $\{z : \text{Arg } z = c\}$, where $-\pi < c \leq \pi$, in the z-plane to height c, as shown in Figure 2.2(b) for $c = -\pi/4$ and $c = \pi/3$. Note that since $\text{Arg } z$ is not defined for $z = 0$, no point on the s-axis is part of the surface. Figure 2.2(a) shows the lifted rays for $c = -\pi/2$, $-\pi/4$, 0, $\pi/6$, $\pi/3$, $\pi/2$ and π lying in the surface $s = \text{Arg } z$. Each lifted ray is the intersection of the surface with the plane $s = c$.

Sketching such surfaces is rather demanding from an artistic point of view, and we do not expect you to be skilled at it. However, it will help you to try the following problem, but do not spend more than a few minutes on it.

Problem 2.1

Sketch the following surfaces.

(a) $s = \text{Re } z$ (b) $s = \text{Im } z$

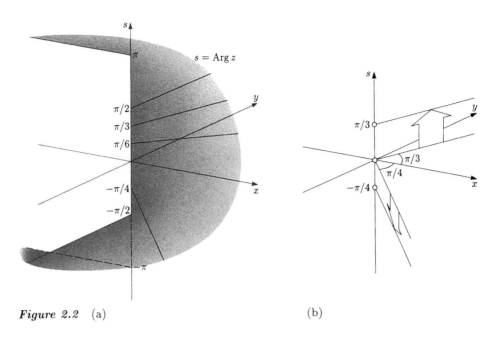

Figure 2.2 (a)   (b)

Real-valued functions of a complex variable arise naturally when we study complex functions. For example, if $f(z) = z^2$, where $z = x + iy$, then

$$f(z) = (x+iy)^2 = (x^2 - y^2) + 2ixy;$$

thus

$$f(z) = u + iv,$$

where

$$u = x^2 - y^2 \quad \text{and} \quad v = 2xy. \tag{2.1}$$

Here both $z \longmapsto u$ and $z \longmapsto v$ are real-valued functions of the complex variable z.

In general, for any complex function f, we can write $f(z)$ in the form

$$f(z) = u + iv,$$

where u and v are real. Thus the functions $z \longmapsto u$ and $z \longmapsto v$ are real-valued functions with the same domain as f. They are called the **real** and **imaginary parts of** f, written Re f and Im f, respectively. Thus, using the notation for the real and imaginary parts of a complex number introduced in *Unit A1*, we have

$$\text{Re } f : z \longmapsto \text{Re}(f(z)) \quad \text{and} \quad \text{Im } f : z \longmapsto \text{Im}(f(z)).$$

In particular,
$$(\text{Re } f)(z) = \text{Re}(f(z)).$$

Problem 2.2

Determine the functions Re f and Im f for the function $f(z) = 1/z$.

If we are given two real-valued functions with the same domain A in $\mathbb{C}$, g and h say, then we can combine them to obtain a function $f : A \longrightarrow \mathbb{C}$ by writing

$$f(z) = g(z) + ih(z) \quad (z \in A).$$

For example, both

$$g(z) = \log_e |z|$$

and

$$h(z) = \text{Arg } z$$

are real-valued functions with domain $\mathbb{C} - \{0\}$ and the function

$$f(z) = g(z) + ih(z)$$
$$= \log_e |z| + i \text{ Arg } z$$

also has domain $\mathbb{C} - \{0\}$.

Figure 2.2(a) gives the graph of the surface $s = \operatorname{Arg} z$, and the graph of the surface $s = \log_e |z|$ is shown in Figure 2.3; for each $c > 0$, the horizontal plane $s = c$ intersects the surface in a circle of radius e^c.

We shall discuss the function $f(z) = \log_e |z| + i \operatorname{Arg} z$ and some of its properties later in the unit.

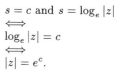

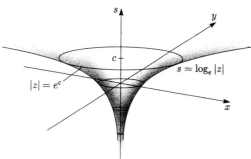

Figure 2.3

2.2 Functions with domain a subset of $\mathbb{R}$

For any complex function f it is possible to sketch the graphs of the two real-valued functions $\operatorname{Re} f$ and $\operatorname{Im} f$. However, it is difficult to gain insight into the geometric nature of the function f from these two surfaces. It would be much more helpful to be able to picture how the image point $w = f(z)$ behaves as z moves around the domain of f. In order to do this, we first need to make precise what it means for the point z to move around the complex plane.

When a point moves in a plane, it traces a *curve* or *path* as time passes. The position of the point on this path can be described by giving both the x- and y-coordinates of the point as functions of time, t. In this context, the real variable t is called a *parameter*. For example, suppose that the x- and y-coordinates are given by the equations

$$x = \cos t, \quad y = \sin t \quad (t \in [0, 2\pi]).$$

Then, as time t increases from 0 to 2π, the point

$$z = x + iy = \cos t + i \sin t$$

moves around the circle Γ with centre 0 and radius 1, starting ($t = 0$) and finishing ($t = 2\pi$) at the point 1, as indicated in Figure 2.4.

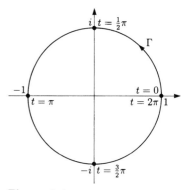

Figure 2.4

If we introduce the function

$$\gamma(t) = \cos t + i \sin t \quad (t \in [0, 2\pi]),$$

then the circle Γ is the image of γ; that is, $\Gamma = \gamma([0, 2\pi])$. The function γ describes a mode of traversing the circle Γ. In general, a set Γ and the associated function γ are the ingredients in our definition of the term 'path'.

Definitions A **path** is a subset Γ of $\mathbb{C}$ which is the image of an associated continuous function $\gamma : I \longrightarrow \mathbb{C}$, where I is a real interval. In this context, the function γ is called a **parametrization** (of (Γ)). If

$$\gamma(t) = \phi(t) + i\psi(t) \quad (t \in I)$$

where ϕ and ψ are real functions, then the equations

$$x = \phi(t), \quad y = \psi(t) \quad (t \in I),$$

are called **parametric equations** (of (Γ)).

Figure 2.5 illustrates these definitions in the case of a closed interval $I = [a, b]$.

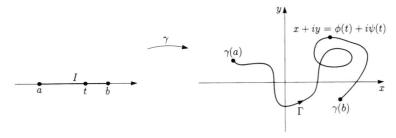

Figure 2.5

Remarks

1 We often speak of 'the path Γ' without referring specifically to the associated parametrization γ. Sometimes it is convenient to speak of 'the path $\Gamma : \gamma$' or 'the path $\Gamma : \gamma(t) = \ldots$'.

2 The condition that the function $\gamma : I \longrightarrow \mathbb{C}$ be continuous is included to ensure that the path Γ has no gaps in it. We shall define the notion of continuity precisely in *Unit A3* but, meanwhile, we point out that all functions γ considered in this context *are* continuous.

3 In Figure 2.5, the interval I is closed: $I = [a, b]$. In this case, the points $\gamma(a)$ and $\gamma(b)$ are called the **initial point** and **final point** respectively of the path. For example, if

$$\gamma(t) = \cos t + i \sin t \qquad (t \in [0, 2\pi]),$$

then the initial point $\gamma(0)$ and the final point $\gamma(2\pi)$ both equal 1.

As in Figure 2.5, a path is usually marked with an arrow (or arrows, if necessary) to show the direction in which it is traversed.

4 It is often possible to eliminate the parameter t from the parametric equations to obtain the equation of the path in terms of x and y alone. For example, if

$$x = \cos t, \quad y = \sin t \qquad (t \in [0, 2\pi]),$$

then

$$x^2 + y^2 = \cos^2 t + \sin^2 t = 1.$$

However, unlike the parametric equations, the equation $x^2 + y^2 = 1$ does not tell us, for example, which way the arrow goes on the path.

5 It is useful to plot a few points of the path to help us to understand the shape of a given path. This is done in the following example.

Example 2.1

Let

$$\gamma(t) = t^2 + it^3 \qquad (t \in \mathbb{R}).$$

Plot the points $\gamma(-1)$, $\gamma\left(-\tfrac{1}{2}\right)$, $\gamma(0)$, $\gamma\left(\tfrac{1}{2}\right)$, $\gamma(1)$ and hence sketch the path Γ with parametrization γ. Determine the equation of the path Γ in terms of x and y.

Solution

First we compile a table of the required values of $x = t^2$ and $y = t^3$.

t	-1	$-\frac{1}{2}$	0	$\frac{1}{2}$	1
x	1	$\frac{1}{4}$	0	$\frac{1}{4}$	1
y	-1	$-\frac{1}{8}$	0	$\frac{1}{8}$	1

We plot the points (x, y) and hence sketch the path Γ (Figure 2.6).

In order to eliminate t from the parametric equations $x = t^2$ and $y = t^3$, we note that
$$x^3 = (t^2)^3 = t^6 = (t^3)^2 = y^2,$$
so that Γ has the equation $y^2 = x^3$. ∎

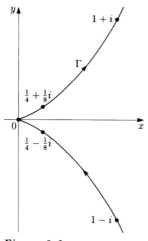

Figure 2.6
The top half of Γ has equation
$$y = x^{3/2};$$
whereas the bottom half has equation
$$y = -x^{3/2}.$$

Problem 2.3

Sketch the paths Γ with the following parametrizations.

(a) $\gamma(t) = 1 + it$ $(t \in \mathbb{R})$ (b) $\gamma(t) = t^2 + it$ $(t \in [-1, 1])$
(c) $\gamma(t) = 1 - t + it$ $(t \in [0, 1])$ (d) $\gamma(t) = 2\cos t + 5i \sin t$ $(t \in [0, 2\pi])$

In each case determine the equation of Γ in terms of x and y.

It is important to realize that a given set can be considered as many different paths by using different parametrizations. For example, the functions
$$\gamma(t) = t \quad (t \in [0, 1])$$
and
$$\gamma(t) = t^2 \quad (t \in [0, 1])$$
are both parametrizations of the real interval $[0, 1]$ in $\mathbb{C}$.

As indicated in Figure 2.7, for the parametrization $\gamma(t) = t$, the progress of the point along the interval $[0, 1]$ is uniform — for example, it is halfway along at time $t = \frac{1}{2}$. But for the parametrization $\gamma(t) = t^2$, the speed of the point varies with t — for example, in the time interval $0 \leq t \leq \frac{1}{2}$, the point has travelled one-quarter of the distance along the interval $[0, 1]$, and in the time interval $\frac{1}{2} \leq t \leq 1$, it travels the next three-quarters. (In fact, in *Unit A4*, you will see that the speed of the point at time t is given by the modulus of the derivative of γ at t.)

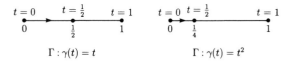

Figure 2.7

Various types of sets (such as line segments, arcs of circles) occur frequently in the course as paths. We shall normally use a standard parametrization for each of these, as indicated in the table opposite.

Problem 2.4

For each of the following paths, write down the standard parametrization and obtain the corresponding parametric equations.

(a) The line through -2 and i.   (b) The line segment between 1 and $1 + i$.
(c) The circle with centre $1 + i$ and radius 1.   (d) The parabola $y = x^2$.

Set	Standard parametrization	Diagram
Line through points α and β.	$\gamma(t) = (1-t)\alpha + t\beta \quad (t \in \mathbb{R})$	
Line segment between points α and β.	$\gamma(t) = (1-t)\alpha + t\beta \quad (t \in [0,1])$	
Circle with centre α, radius r: $|z - \alpha| = r$.	$\gamma(t) = \alpha + r(\cos t + i\sin t) \quad (t \in [0, 2\pi])$	
Arc of circle with centre α, radius r.	$\gamma(t) = \alpha + r(\cos t + i\sin t) \quad (t \in [t_1, t_2])$	
Ellipse in standard form: $\dfrac{x^2}{a^2} + \dfrac{y^2}{b^2} = 1.$	$\gamma(t) = a\cos t + ib\sin t \quad (t \in [0, 2\pi])$	
Parabola in standard form: $y^2 = 4ax.$	$\gamma(t) = at^2 + 2iat \quad (t \in \mathbb{R})$	
Right half of hyperbola in standard form: $\dfrac{x^2}{a^2} - \dfrac{y^2}{b^2} = 1.$	$\gamma(t) = a\cosh t + ib\sinh t \quad (t \in \mathbb{R})$	

Now we consider the following question. How does the image point $w = f(z)$ behave as the point z moves around the domain of the function f? More precisely, as z moves along a path Γ in the domain, what is the set of image points

$$\{f(z) : z \in \Gamma\}?$$

We call this set the *image* under f of Γ. We make the following definition.

Definition Given a function $f : A \longrightarrow B$ and a subset S of A, the **image under f of S**, written $f(S)$, is
$$f(S) = \{f(z) : z \in S\}.$$

Remark If $S = A$, then, as noted earlier, $f(S)$ is also described as the image of f.

Suppose that $f(z) = z^2$ and that the point z moves around the unit circle $|z| = 1$. In this case, our knowledge of the function f tells us that the image w of z satisfies $|w| = |z^2| = |z|^2 = 1$, and

if θ is an argument of z, then 2θ is an argument of w (Figure 2.8).

Thus, if z moves once around the circle $|z| = 1$ anticlockwise, then w moves twice around the circle $|w| = 1$ anticlockwise.

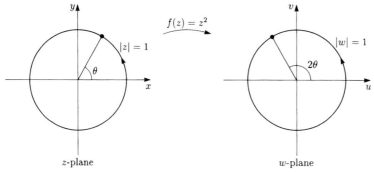

Figure 2.9

This geometric observation can be made precise by using the parametrization
$$\gamma(t) = \cos t + i \sin t \quad (t \in [0, 2\pi]),$$
of the unit circle $|z| = 1$ traversed once anticlockwise. Now from Equations (2.1) we know that for $f(z) = z^2$, with $z = x + iy$ and $w = f(z) = u + iv$,
$$u = x^2 - y^2, \quad v = 2xy. \tag{2.2}$$

The parametric equations corresponding to the parametrization γ are
$$x = \cos t, \quad y = \sin t \quad (t \in [0, 2\pi]) \tag{2.3}$$
and, on substituting these in Equations (2.2), we obtain
$$u = \cos^2 t - \sin^2 t, \quad v = 2 \cos t \sin t;$$
that is,
$$u = \cos 2t, \quad v = \sin 2t \quad (t \in [0, 2\pi]).$$

These are the parametric equations for the image circle $|w| = 1$ corresponding to Equations (2.3). Thus, as t increases from 0 to 2π, z moves anticlockwise *once* round the circle $|z| = 1$, whereas the image $w = f(z)$ moves anticlockwise *twice* round the circle $|w| = 1$.

Alternatively, we have
$$u + iv = (x + iy)^2$$
$$= (\cos t + i \sin t)^2$$
$$= \cos 2t + i \sin 2t,$$
by de Moivre's Theorem.

In the next section, we systematically investigate the images of some particular types of path for a number of functions. We use the above approaches, which are summarized in the following definition and strategy.

If f is a continuous function and Γ is a path in the domain of f, then $f(\Gamma)$ is called the **image path**. If Γ has parametrization γ, then $f(\Gamma)$ has parametrization $f \circ \gamma$. (The continuity of $f \circ \gamma$ follows from the Composition Rule, which is established in *Unit A3*.)

$f \circ \gamma$ is the function with rule $t \longmapsto f(\gamma(t))$.

Strategy for determining an image path

Let the parametrization of the path Γ be
$$\gamma(t) = \phi(t) + i\psi(t) \quad (t \in I).$$
Then the image path $f(\Gamma)$ is found

EITHER

by using the geometric properties of the continuous function f

OR

by substituting $x = \phi(t)$, $y = \psi(t)$ into the equation
$$u + iv = f(x + iy)$$
and then, by equating real parts and imaginary parts, obtaining expressions for u and v in terms of t. (These expressions are the parametric equations of the image path $f(\Gamma)$, associated with the parametrization $f \circ \gamma$.)

---

3 IMAGES OF GRIDS (AUDIO-TAPE)

After working through this section, you should be able to:

(a) sketch the image of a given Cartesian grid under various complex functions;

(b) sketch the image of a given polar grid under various complex functions.

To obtain a clear picture of the geometric nature of a given complex function, we consider the images of many lines or paths in the domain. In order to do this in a systematic way we introduce two types of grid. The first type is a **Cartesian grid** consisting of lines of the form $x = a$ and $y = b$, usually evenly spaced in each direction (Figure 3.1).

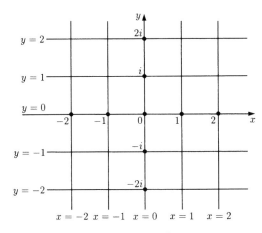

Figure 3.1 A Cartesian grid

Each of the lines in such a grid can be thought of as a path, but to avoid cluttering diagrams we shall usually omit the corresponding arrows.

The second type of grid consists of circles with centre 0 and rays emerging from 0; it is called a **polar grid** because of the connection with the polar form $z = r(\cos\theta + i\sin\theta)$. Each of the circles has an equation of the form $r = a$, where a is a positive constant, and each of the rays has an equation of the form $\theta = b$, where b is a constant in the interval $]-\pi, \pi]$ (Figure 3.2).

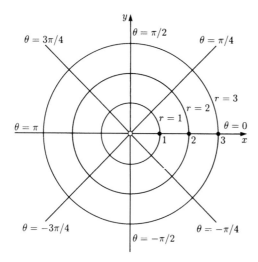

Figure 3.2 A polar grid

Problem 3.1

Plot the polar grid consisting of the circles $r = 1$, $r = \frac{1}{2}$, $r = \frac{1}{3}$ and the rays $\theta = 0$, $\theta = \pm\pi/3$, $\theta = \pm 2\pi/3$, $\theta = \pi$.

In the audio tape we consider several complex functions f. In each case we sketch the images of a Cartesian grid and a polar grid, in order to obtain a clear picture of the geometric effect of f. These images are found by using the strategy given at the end of Subsection 2.2.

In each case we have also highlighted (by using tone) the effect of the function on a particular set bounded by parts of the grid. (Note that in some frames the scales in the z-plane and w-plane are not the same.)

Now try the following problem, the results of which you will use when working through the audio tape.

Problem 3.2

Eliminate t from each of the following pairs of parametric equations (in each case, a is a real constant, with $a \neq 0$ in parts (b) and (c)).

(a) $u = a - t$, $v = a + t$ (b) $u = a^2 - t^2$, $v = 2at$

(c) $u = \dfrac{a}{a^2 + t^2}$, $v = \dfrac{-t}{a^2 + t^2}$

NOW START THE TAPE.

1. The function $f(z) = (1+i)z$

If $z = x + iy$ and $w = u + iv$, then $w = (1+i)z$ gives
$$u + iv = (1+i)(x+iy)$$
$$= (x-y) + i(x+y).$$

Thus
$$u = x-y \quad \text{and} \quad v = x+y.$$

To find the image of the line $x = a$:
$x = a, y = t$ gives $u = a - t, v = a + t$.

Eliminate t:
$$u + v = 2a, \text{ a line.}$$

Parametric equations: $t \in \mathbb{R}$

Put $w = f(z)$.

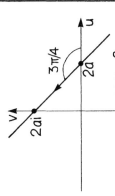

2. Problem 3.3

Determine the image of the line
$$y = b,$$
where $b \in \mathbb{R}$, under the function $f(z) = (1+i)z$.

Sketch the images of the lines $y = 1$ and $y = 0$.

3. Image of Cartesian grid

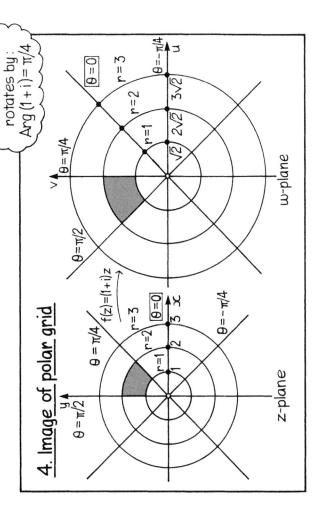

scales by: $|1+i| = \sqrt{2}$
rotates by: $\text{Arg}(1+i) = \pi/4$

4. Image of polar grid

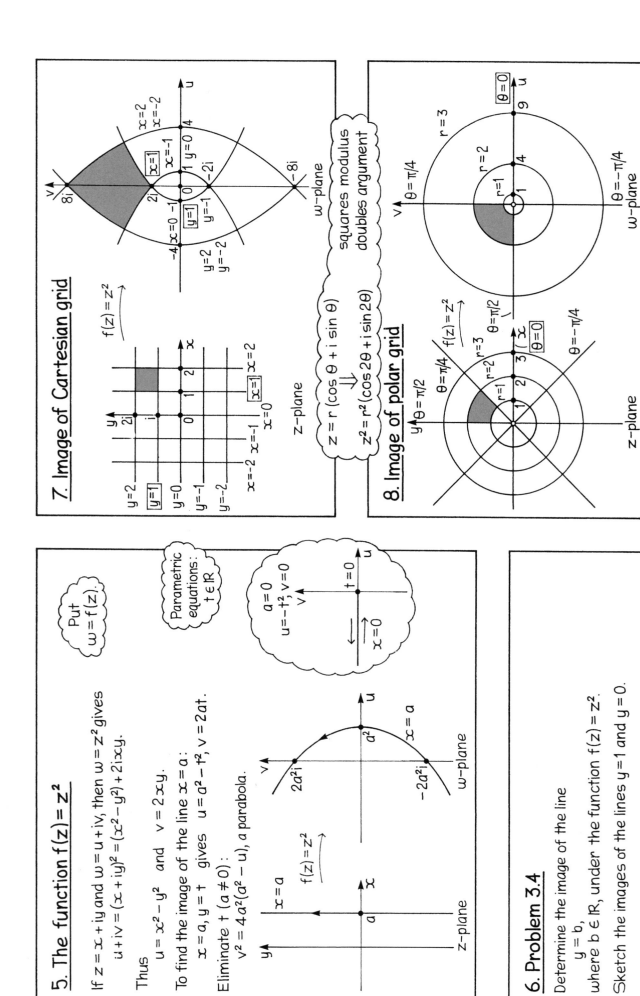

9. The function $f(z) = 1/z$

If $z = x+iy$ and $w = u+iv$, then $w = 1/z$ gives

$$u + iv = \frac{1}{x+iy} = \frac{x-iy}{(x+iy)(x-iy)} = \frac{x-iy}{x^2+y^2}.$$

Thus

$$u = \frac{x}{x^2+y^2} \quad \text{and} \quad v = \frac{-y}{x^2+y^2}.$$

To find the image of the line $x = a$:

$x = a, y = t$ gives $u = \dfrac{a}{a^2+t^2}$, $v = \dfrac{-t}{a^2+t^2}$

Eliminate t ($a \neq 0$):

$u^2 + v^2 = \dfrac{u}{a}$, a circle through $(0,0)$.

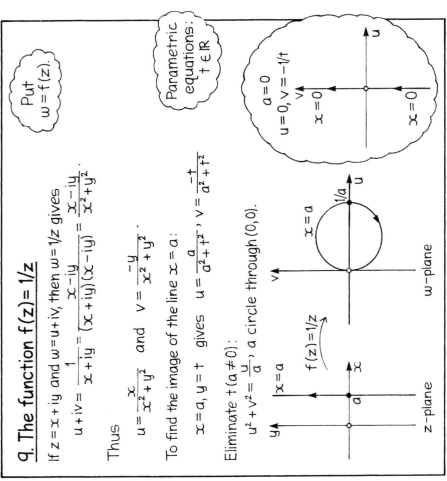

Put $w = f(z)$.

Parametric equations: $t \in \mathbb{R}$

$a = 0$
$u = 0, v = -1/t$
$x = 0$
$x = 0$

10. Problem 3.5

Determine the image of the line
$y = b$,
where $b \in \mathbb{R}$, under the function $f(z) = 1/z$.

Sketch the images of the lines $y = 1$ and $y = 0$.

11. Image of Cartesian grid

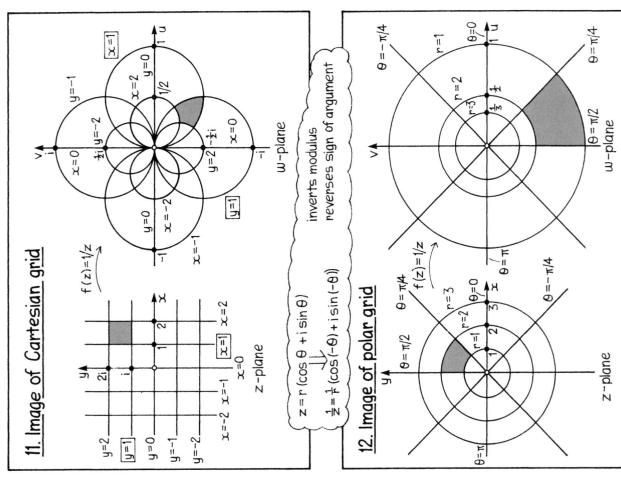

inverts modulus
reverses sign of argument

$z = r(\cos\theta + i\sin\theta)$
$\dfrac{1}{z} = \dfrac{1}{r}(\cos(-\theta) + i\sin(-\theta))$

12. Image of polar grid

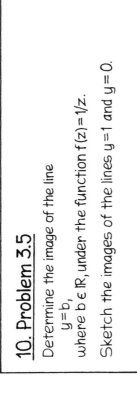

In the audio tape we determined the images of a Cartesian grid and a polar grid for each of several complex functions. The following problems provide some practice in the use of these techniques. In such problems your geometric knowledge of the function f may save your getting involved in parametrizations.

Problem 3.6

For the function $f(z) = iz + 1$, sketch the images of

(a) $S = \{z : 0 \leq \operatorname{Re} z \leq 1, 0 \leq \operatorname{Im} z \leq 1\}$;

(b) the polar grid of Figure 3.2, which was used in the audio-tape frames.

Problem 3.7

For the function $f(z) = z^3$, sketch the images of

(a) $S = \{z : 0 \leq \operatorname{Re} z \leq 1, 0 \leq \operatorname{Im} z \leq 1\}$;

(b) the polar grid of Figure 3.2, with the circle with equation $r = 3$ omitted (its image is rather large).

Problem 3.8

Find the image of the polar grid of Figure 3.2 under the function

$$f(z) = \sqrt{z}.$$

4 EXPONENTIAL, TRIGONOMETRIC AND HYPERBOLIC FUNCTIONS

After working through this section, you should be able to:

(a) state and use the definition and basic algebraic properties of the exponential function and describe its geometric properties;

(b) state and use the definitions and basic algebraic properties of the trigonometric functions and hyperbolic functions;

(c) prove simple identities involving the exponential function, the trigonometric and hyperbolic functions.

4.1 The exponential function

In Section 3 we considered rational functions, which can be evaluated using the four basic operations of addition, subtraction, multiplication and division. The functions introduced in this section cannot be evaluated in this way.

The real exponential function $x \longmapsto e^x$ (Figure 4.1) arises frequently in mathematics, in particular in the solution of differential equations. It is natural to ask whether this function has a complex analogue; that is, how is e^z defined where z is a complex number?

A fundamental property of the real exponential function is that

$$e^{x_1} e^{x_2} = e^{x_1 + x_2}, \quad \text{for all } x_1, x_2 \in \mathbb{R}.$$

Ideally then, the complex exponential function should satisfy

$$e^{z_1} e^{z_2} = e^{z_1 + z_2}, \quad \text{for all } z_1, z_2 \in \mathbb{C}. \tag{4.1}$$

In particular, if $z = x + iy$, then it should be true that

$$e^z = e^{x+iy} = e^x e^{iy}.$$

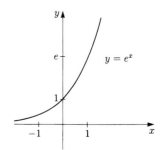

Figure 4.1

Since e^x is defined, because x is real, it remains only to define e^{iy}, where y is real. The following manipulation of power series suggests a definition of e^{iy}. The power series for e^x, where $x \in \mathbb{R}$, is

$$e^x = 1 + x + \frac{x^2}{2!} + \frac{x^3}{3!} + \cdots \qquad (x \in \mathbb{R}).$$

If we replace x by iy, then we obtain

$$e^{iy} = 1 + iy + \frac{(iy)^2}{2!} + \frac{(iy)^3}{3!} + \cdots$$

$$= \left(1 - \frac{y^2}{2!} + \cdots\right) + i\left(y - \frac{y^3}{3!} + \cdots\right).$$

We cannot justify replacing x by iy at this stage, since iy is not a real number.

The expressions in parentheses are the power series for $\cos y$ and $\sin y$, respectively. Thus it seems plausible to define e^{iy} by

$$e^{iy} = \cos y + i \sin y;$$

this formula is known as **Euler's Identity**, although it was known to others (for example, to de Moivre) before Euler.

Definition   For all $z = x + iy$ in $\mathbb{C}$,
$$e^z = e^x(\cos y + i \sin y).$$
The function
$$z \longmapsto e^z \qquad (z \in \mathbb{C})$$
is called the **exponential function**, and is denoted by exp.

Thus $\exp z = e^z$.

Before verifying that this definition does indeed satisfy Equation (4.1), we evaluate e^z for several complex numbers z and then make some remarks concerning the definition.

Example 4.1
Express each of the following numbers in Cartesian form.
(a) $e^{i\pi/3}$ (b) $e^{(1-i\pi)/2}$ (c) $e^{-1+i\pi/4}$

Solution
(a) $e^{i\pi/3} = e^0(\cos \pi/3 + i \sin \pi/3) = \frac{1}{2} + \frac{\sqrt{3}}{2}i$
(b) $e^{(1-i\pi)/2} = e^{1/2}(\cos(-\pi/2) + i \sin(-\pi/2)) = -ie^{1/2}$
(c) $e^{-1+i\pi/4} = e^{-1}(\cos \pi/4 + i \sin \pi/4) = e^{-1}\left(\frac{1}{\sqrt{2}} + \frac{1}{\sqrt{2}}i\right)$ ∎

Remarks
1 Notice that if z is real, so that $z = x + 0i$ (that is, $z = x$), then
$$e^z = e^x(\cos 0 + i \sin 0) = e^x.$$

Thus, the restriction of the (complex) exponential function to $\mathbb{R}$ gives the real exponential function, as we would expect. In particular, for $0 \in \mathbb{C}$,
$$e^0 = 1.$$

2 Also, if z is imaginary, so that $z = 0 + iy$, then
$$e^z = e^0(\cos y + i \sin y),$$
which gives Euler's Identity
$$e^{iy} = \cos y + i \sin y.$$

Thus, the number e^{iy} has modulus 1 and argument y, and so lies on the unit circle $\{z : |z| = 1\}$. Some important examples are given in Figure 4.2.

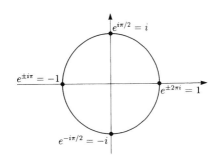

Figure 4.2

In particular, $e^{i\pi} = -1$; that is,
$$e^{i\pi} + 1 = 0.$$

This striking equation contains the five most important numbers in mathematics, together with two of the most important symbols.

Problem 4.1

Express each of the following numbers in Cartesian form.

(a) $e^{2\pi i}$ (b) $e^{2+i\pi/3}$ (c) $e^{-(1+i\pi)}$

The next result gives a number of basic identities involving the exponential function, including Equation (4.1). Here and subsequently, we adopt the convention that, unless otherwise stated, identities hold for all values of the variables for which the identity has meaning. For example, the first identity below holds for all z_1 and z_2 in $\mathbb{C}$.

Convention on variables in identities.

Theorem 4.1 Exponential Identities

(a) **Addition**
$$e^{z_1+z_2} = e^{z_1}e^{z_2}.$$

(b) **Modulus**
$$|e^z| = e^{\operatorname{Re} z}.$$

(c) **Negatives**
$$e^{-z} = 1/e^z.$$

(d) **Periodicity**
$$e^{z+2\pi i} = e^z.$$

Remarks

1 One consequence of part (a) is that
$$(e^{i\theta})^n = e^{in\theta}, \quad \text{for } \theta \in \mathbb{R}, n \in \mathbb{Z}. \tag{4.2}$$
This is a restatement of de Moivre's Theorem in a concise form.

Theorem 2.2, Unit A1.

2 Since the real exponential function is always positive, part (b) shows that $e^z \neq 0$ for all $z \in \mathbb{C}$.

3 Part (d) shows that the exponential function is periodic with period $2\pi i$:
$$\exp(z + 2\pi i) = \exp z.$$

(In fact, it is easy to show that any other period is an integer multiple of $2\pi i$.)

Proof We prove only parts (a), (b) and (c), leaving part (d) as a problem.

(a) Let $z_1 = x_1 + iy_1$ and $z_2 = x_2 + iy_2$. Then
$$\begin{aligned}
e^{z_1}e^{z_2} &= e^{x_1}(\cos y_1 + i \sin y_1)e^{x_2}(\cos y_2 + i \sin y_2) \\
&= e^{x_1}e^{x_2}(\cos y_1 + i \sin y_1)(\cos y_2 + i \sin y_2) \\
&= e^{x_1+x_2}((\cos y_1 \cos y_2 - \sin y_1 \sin y_2) \\
&\qquad + i(\sin y_1 \cos y_2 + \cos y_1 \sin y_2)) \\
&= e^{x_1+x_2}(\cos(y_1 + y_2) + i \sin(y_1 + y_2)) \\
&= e^{z_1+z_2}, \quad \text{since } z_1 + z_2 = (x_1 + x_2) + i(y_1 + y_2).
\end{aligned}$$

(b) Let $z = x + iy$. Then
$$\begin{aligned}
|e^z| &= |e^x(\cos y + i \sin y)| \\
&= |e^x|\,|\cos y + i \sin y| \\
&= e^x, \quad \text{since } e^x > 0 \text{ and } \cos^2 y + \sin^2 y = 1, \\
&= e^{\operatorname{Re} z}.
\end{aligned}$$

(c) By part (a),
$$\begin{aligned}
e^z e^{-z} &= e^{z-z} \\
&= e^0 = 1,
\end{aligned}$$
and hence $e^{-z} = 1/e^z$. (From Remark 2, $e^z \neq 0$ for all $z \in \mathbb{C}$.) ∎

Problem 4.2

(a) Prove part (d) of Theorem 4.1.

(b) Prove that
$$|e^z| \le e^{|z|}, \quad \text{for all } z \in \mathbb{C}.$$

The result in part (b) will be useful later in the course.

(c) Is the function exp one-one?

(d) Determine each of the following sets.
 (i) $\{z \in \mathbb{C} : e^z = 1\}$ (ii) $\{z \in \mathbb{C} : e^z = -1\}$

The exponential function provides an alternative and very concise notation for expressing a non-zero complex number in polar form. For example,
$$1 + i = \sqrt{2}(\cos \pi/4 + i \sin \pi/4) = \sqrt{2}\, e^{i\pi/4}.$$

Thus, it is easy to compute powers such as $(1+i)^{10}$ as follows:

$$\begin{aligned}(1+i)^{10} &= (\sqrt{2}\, e^{i\pi/4})^{10} \\ &= (\sqrt{2})^{10}(e^{i\pi/4})^{10} \\ &= 32 e^{10i\pi/4} \quad \text{(by Equation (4.2))} \\ &= 32 e^{5i\pi/2} \\ &= 32 e^{i\pi/2} \quad \text{(exp has period } 2\pi i\text{)} \\ &= 32i.\end{aligned}$$

Compare this calculation with that in Problem 2.12(c) of Unit A1.

Problem 4.3

Use exponential notation to evaluate $(\sqrt{3} + i)^{-6}$.

The geometric nature of the exponential function

Part (d) of Theorem 4.1 shows that the (complex) exponential function is not one-one, since
$$e^{z+2\pi i} = e^z \tag{4.3}$$
but $z + 2\pi i \ne z$. Repeated application of Equation (4.3) shows that
$$e^{z+2n\pi i} = e^z, \quad \text{for } n \in \mathbb{Z},$$
so that each of the points
$$z + 2n\pi i, \quad n \in \mathbb{Z},$$
has the same image under the exponential function. This property has a bearing on the geometric nature of the exponential function, which we now investigate.

As in Section 3, the aim is to plot the image of a grid of lines of the form $x = a$ and $y = b$, for suitable constants a and b. To do this let $w = e^z$, where $z = x + iy$ and $w = u + iv$, so that
$$u + iv = e^z = e^x(\cos y + i \sin y).$$
Hence
$$u = e^x \cos y \quad \text{and} \quad v = e^x \sin y. \tag{4.4}$$

First we consider the image of a line of the form $x = a$, which has the parametrization $\gamma(t) = a + it$ ($t \in \mathbb{R}$). Substituting $x = a$ and $y = t$ into Equations (4.4), we obtain the following property.

As shown in Figure 4.3, each of these points lies on the vertical line through z.

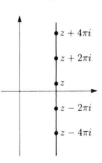

Figure 4.3

The function exp maps the line $x = a$ to the path with parametric equations
$$u = e^a \cos t, \quad v = e^a \sin t \quad (t \in \mathbb{R}).$$
This is the circle with centre 0 and radius e^a (Figure 4.4).

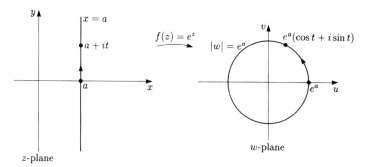

Figure 4.4 The line $x = a$ and its image circle $\{w : |w| = e^a\}$

Notice that

(a) as t increases, the image point w moves anticlockwise around the image circle, passing through e^a whenever t is an integer multiple of 2π;

(b) the image of the line $x = 0$ is the unit circle $\{w : |w| = 1\}$;

(c) as a increases, the image circle of the line $x = a$ expands, the centre remaining fixed at 0.

Next consider the image of a line of the form $y = b$, which has the parametrization $\gamma(t) = t + ib$ ($t \in \mathbb{R}$). Substituting $x = t$ and $y = b$ into Equations (4.4), we obtain the following property.

The function exp maps the line $y = b$ to the path with parametric equations
$$u = e^t \cos b, \quad v = e^t \sin b \quad (t \in \mathbb{R}).$$
This is the ray from 0 (excluded) through $\cos b + i \sin b$ (Figure 4.5).

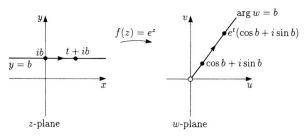

Figure 4.5 The line $y = b$ and its image ray $\{w : \arg w = b\}$

Notice that

(a) as t increases, the image point w moves outwards along the image ray;

(b) the image of the line $y = 0$ is the positive real axis;

(c) as b increases, the image ray of the line $y = b$ rotates anticlockwise about 0.

Combining these observations, we can now plot the image of a grid of lines of the form $x = a$ and $y = b$. For our grid we choose the values of a to be integers (as usual) but, because trigonometric functions are involved, it is convenient to choose the values of b to be integer multiples of $\pi/2$.

In Figure 4.6, the image circles of the lines $x = -2, -1, 0, 1, 2$ are shown, as are the image rays of the lines $y = -3\pi/2, -\pi, -\pi/2, 0, \pi/2, \pi$. (Note that the lines $y = -\pi$ and $y = \pi$, for example, have the same image.)

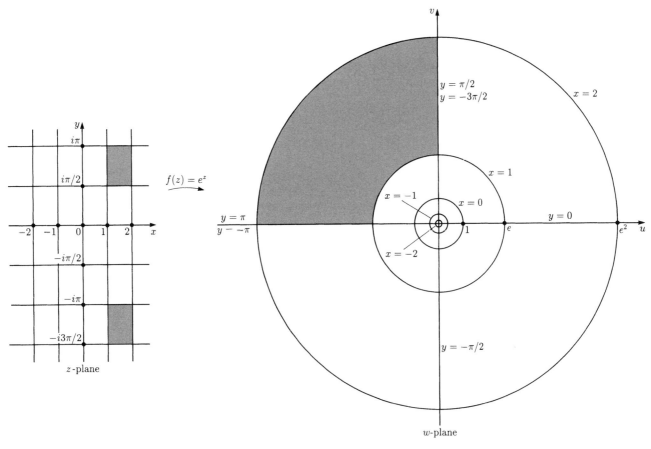

Figure 4.6

One effect of this choice of values for b is that the image of each grid rectangle in the z-plane is a quarter-annulus (sweeping out the angle $\pi/2$) in the w-plane. In particular, notice that the two shaded rectangles in Figure 4.6 map to the same quarter-annulus.

Notice also that

(a) since $|e^z| = e^{\operatorname{Re} z}$, points in the right half-plane $\{z : \operatorname{Re} z > 0\}$ have images lying outside the circle $\{w : |w| = 1\}$, whereas those in the left half-plane $\{z : \operatorname{Re} z < 0\}$ have images lying inside this circle;

(b) the image of the strip $\{x + iy : -\pi < y \leq \pi\}$ is the whole of the complex plane, except for the point 0 (Figure 4.7).

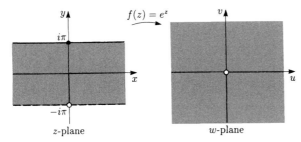

Figure 4.7

This latter property will prove useful in Section 5, where we discuss possible inverse functions for the exponential function.

Problem 4.4

Sketch the image of each of the following sets under the exponential function.

(a) $\{x + iy : -1 \leq x \leq 0, -\pi/4 \leq y \leq \pi/4\}$

(b) $\{x + iy : -1 \leq x \leq 1, \pi \leq y \leq 2\pi\}$

(c) $\{x + iy : 0 < y < 2\pi\}$

4.2 Trigonometric functions

In the study of real functions there seems, at first sight, to be no connection between the trigonometric functions $x \longmapsto \sin x$ and $x \longmapsto \cos x$, with their geometric definitions, and the exponential function $x \longmapsto e^x$. Certainly, their graphs (Figure 4.8) do not suggest any connection.

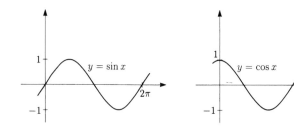

Figure 4.8

However, the definition of the *complex* exponential function makes use of real trigonometric functions and it turns out that this complex exponential function can be used to define complex trigonometric functions.

The key to such a definition is Euler's Identity

$$e^{i\theta} = \cos\theta + i\sin\theta, \qquad \theta \in \mathbb{R},$$

together with the consequence

$$e^{-i\theta} = \cos\theta - i\sin\theta, \qquad \theta \in \mathbb{R}.$$

$\cos(-\theta) = \cos\theta;$
$\sin(-\theta) = -\sin\theta.$

Eliminating first $\sin\theta$ and then $\cos\theta$ from these equations, we obtain

$$\cos\theta = \tfrac{1}{2}(e^{i\theta} + e^{-i\theta}) \quad \text{and} \quad \sin\theta = \frac{1}{2i}(e^{i\theta} - e^{-i\theta}). \qquad (4.5)$$

These equations suggest the following definitions.

Definitions For all z in $\mathbb{C}$,

$$\cos z = \tfrac{1}{2}(e^{iz} + e^{-iz}) \quad \text{and} \quad \sin z = \frac{1}{2i}(e^{iz} - e^{-iz}).$$

The corresponding cosine and sine functions are called cos and sin.

Remark Note that if z is real, then these definitions lead to Equations (4.5).

With these definitions, the functions cos and sin enjoy most (but not all) of the properties of the corresponding real trigonometric functions. Before stating these properties, we determine some values of these functions.

Example 4.2

Evaluate each of the following numbers in Cartesian form.

(a) $\sin i$ (b) $\cos(\pi + i)$

Solution

(a) $$\sin i = \frac{1}{2i}(e^{i^2} - e^{-i^2})$$
$$= \frac{1}{2i}(e^{-1} - e) = \tfrac{1}{2}(e - e^{-1})i$$

(b) $$\cos(\pi + i) = \tfrac{1}{2}(e^{i(\pi+i)} + e^{-i(\pi+i)})$$
$$= \tfrac{1}{2}(e^{i\pi - 1} + e^{-i\pi + 1})$$
$$= \tfrac{1}{2}(e^{i\pi}e^{-1} + e^{-i\pi}e^{1})$$
$$= \tfrac{1}{2}(-e^{-1} - e^{1}) = -\tfrac{1}{2}(e + e^{-1}) \quad \blacksquare$$

Remember that $e^{\pm i\pi} = -1$.

Remarks

1 Notice that
$$|\sin i| = \tfrac{1}{2}(e - e^{-1}) \simeq 1.175 > 1,$$
and
$$|\cos(\pi + i)| = \tfrac{1}{2}(e + e^{-1}) \simeq 1.543 > 1.$$

Thus the well-known properties of the real sine and cosine functions
$$|\sin x| \leq 1, \quad x \in \mathbb{R}, \quad \text{and} \quad |\cos x| \leq 1, \quad x \in \mathbb{R},$$
do not hold for the complex sine and cosine functions.

2 The solutions to Example 4.2 suggest that there is a connection between the complex trigonometric functions and the hyperbolic functions. This will be made clear later in the section.

Problem 4.5

Evaluate each of the following complex numbers in Cartesian form.

(a) $\sin(\pi/2 + i)$ (b) $\cos i$

In order to describe the algebraic properties of the complex sine and cosine functions, we need to introduce the full range of complex trigonometric functions. First, however, we determine the zeros of sin and cos.

Theorem 4.2

(a) The set of zeros of the sine function is
$$\{z : \sin z = 0\} = \{n\pi : n \in \mathbb{Z}\}.$$

(b) The set of zeros of the cosine function is
$$\{z : \cos z = 0\} = \{(n + \tfrac{1}{2})\pi : n \in \mathbb{Z}\}.$$

Thus the only zeros of the complex sine and cosine functions are those of the real sine and cosine functions.

Proof

(a) Using the definition of $\sin z$, we have
$$\sin z = 0 \iff \frac{1}{2i}(e^{iz} - e^{-iz}) = 0$$
$$\iff e^{iz} = e^{-iz}$$
$$\iff e^{2iz} = 1$$
$$\iff 2iz \in \{2n\pi i : n \in \mathbb{Z}\} \quad \text{(see Problem 4.2(d))}$$
$$\iff z \in \{n\pi : n \in \mathbb{Z}\},$$

which proves part (a).

(b) Using the definition of $\cos z$, we have

$$\cos z = 0 \iff \frac{1}{2}(e^{iz} + e^{-iz}) = 0$$
$$\iff e^{iz} = -e^{-iz}$$
$$\iff e^{2iz} = -1$$
$$\iff 2iz \in \{(2n+1)\pi i : n \in \mathbb{Z}\} \quad \text{(see Problem 4.2(d))}$$
$$\iff z \in \{(n + \tfrac{1}{2})\pi : n \in \mathbb{Z}\},$$

which proves part (b). ∎

The other complex trigonometric functions tan, sec, cot and cosec are defined as in the real case.

Definitions For all z in $\mathbb{C} - \{(n + \tfrac{1}{2})\pi : n \in \mathbb{Z}\}$,

$$\tan z = \frac{\sin z}{\cos z} \quad \text{and} \quad \sec z = \frac{1}{\cos z}.$$

For all z in $\mathbb{C} - \{n\pi : n \in \mathbb{Z}\}$,

$$\cot z = \frac{\cos z}{\sin z} \quad \text{and} \quad \operatorname{cosec} z = \frac{1}{\sin z}.$$

The corresponding trigonometric functions are called tan, sec, cot and cosec.

We now record the basic algebraic identities satisfied by these complex trigonometric functions.

Theorem 4.3 Trigonometric Identities

(a) **Addition**

$$\sin(z_1 + z_2) = \sin z_1 \cos z_2 + \cos z_1 \sin z_2;$$
$$\cos(z_1 + z_2) = \cos z_1 \cos z_2 - \sin z_1 \sin z_2;$$
$$\tan(z_1 + z_2) = \frac{\tan z_1 + \tan z_2}{1 - \tan z_1 \tan z_2}.$$

(b) **Squares**

$$\cos^2 z + \sin^2 z = 1; \quad \sec^2 z = 1 + \tan^2 z; \quad \operatorname{cosec}^2 z = 1 + \cot^2 z.$$

(c) **Negatives**

$$\sin(-z) = -\sin z; \quad \cos(-z) = \cos z; \quad \tan(-z) = -\tan z.$$

(d) **Periodicity**

$$\sin(z + 2\pi) = \sin z; \quad \cos(z + 2\pi) = \cos z; \quad \tan(z + \pi) = \tan z.$$

We prove three of these identities in the next example, and ask you to check some more of them in Problem 4.6.

Example 4.3

Prove the following.

(a) $\sin(z_1 + z_2) = \sin z_1 \cos z_2 + \cos z_1 \sin z_2$
(b) $\cos(-z) = \cos z$
(c) $\sin(z + 2\pi) = \sin z$

Solution

(a) Starting with the right-hand side, we have

$$\sin z_1 \cos z_2 + \cos z_1 \sin z_2$$
$$= \left(\frac{e^{iz_1} - e^{-iz_1}}{2i}\right)\left(\frac{e^{iz_2} + e^{-iz_2}}{2}\right) + \left(\frac{e^{iz_1} + e^{-iz_1}}{2}\right)\left(\frac{e^{iz_2} - e^{-iz_2}}{2i}\right)$$
$$= \frac{1}{4i}\{(e^{i(z_1+z_2)} + e^{i(z_1-z_2)} - e^{-i(z_1-z_2)} - e^{-i(z_1+z_2)})$$
$$+ (e^{i(z_1+z_2)} - e^{i(z_1-z_2)} + e^{-i(z_1-z_2)} - e^{-i(z_1+z_2)})\}$$
$$= \frac{1}{2i}(e^{i(z_1+z_2)} - e^{-i(z_1+z_2)})$$
$$= \sin(z_1 + z_2),$$

as required.

(b) We have

$$\cos(-z) = \tfrac{1}{2}(e^{i(-z)} + e^{-i(-z)})$$
$$= \tfrac{1}{2}(e^{iz} + e^{-iz}) = \cos z,$$

as required.

(c) We have

$$\sin(z + 2\pi) = \frac{1}{2i}(e^{i(z+2\pi)} - e^{-i(z+2\pi)})$$
$$= \frac{1}{2i}(e^{iz}e^{2\pi i} - e^{-iz}e^{-2\pi i})$$
$$= \frac{1}{2i}(e^{iz} - e^{-iz}) \qquad\qquad e^{2\pi i} = e^{-2\pi i} = 1$$
$$= \sin z,$$

as required. ∎

Theorem 4.3 is by no means an exhaustive list of trigonometric identities. For example, we have not included identities such as

$$\sin(z_1 - z_2) = \sin z_1 \cos z_2 - \cos z_1 \sin z_2,$$

and

$$\sin 2z = 2 \sin z \cos z.$$

However, these can readily be deduced from the identities in Theorem 4.3.

Problem 4.6

(a) Prove the following.

   (i) $\sin(-z) = -\sin z$ (ii) $\cos(z + 2\pi) = \cos z$

(b) Deduce the following from Theorem 4.3.

   (i) $\cos 2z = 2\cos^2 z - 1$ (ii) $\tan(z_1 - z_2) = \dfrac{\tan z_1 - \tan z_2}{1 + \tan z_1 \tan z_2}$

4.3 Hyperbolic functions

Earlier in this section we referred to a relationship between complex trigonometric functions and the real hyperbolic functions

$$\sinh x = \tfrac{1}{2}(e^x - e^{-x}) \quad \text{and} \quad \cosh x = \tfrac{1}{2}(e^x + e^{-x}),$$

whose graphs appear in Figure 4.9.

The complex hyperbolic functions are defined by the same formulas as for the real hyperbolic functions.

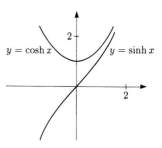

Figure 4.9

Definitions For all z in $\mathbb{C}$,
$$\sinh z = \tfrac{1}{2}(e^z - e^{-z}) \quad \text{and} \quad \cosh z = \tfrac{1}{2}(e^z + e^{-z}).$$
For all z in $\mathbb{C} - \{(n + \tfrac{1}{2})\pi i : n \in \mathbb{Z}\}$,
$$\tanh z = \frac{\sinh z}{\cosh z} \quad \text{and} \quad \operatorname{sech} z = \frac{1}{\cosh z}.$$
For all z in $\mathbb{C} - \{n\pi i : n \in \mathbb{Z}\}$,
$$\coth z = \frac{\cosh z}{\sinh z} \quad \text{and} \quad \operatorname{cosech} z = \frac{1}{\sinh z}.$$
The corresponding hyperbolic functions are called sinh, cosh, tanh, sech, coth, cosech.

In these definitions we have used the facts that
$$\{z : \sinh z = 0\} = \{n\pi i : n \in \mathbb{Z}\}$$
and
$$\{z : \cosh z = 0\} = \{(n + \tfrac{1}{2})\pi i : n \in \mathbb{Z}\}.$$

All these zeros lie on the imaginary axis.

These 'zero sets' are readily deduced from the zero sets of the sine and cosine functions, by using the following result, which shows the very close relationship between the complex hyperbolic functions and the complex trigonometric functions.

Theorem 4.4 For all z in $\mathbb{C}$,
$$\sin(iz) = i \sinh z \quad \text{and} \quad \cos(iz) = \cosh z.$$

Proof For $z \in \mathbb{C}$,
$$\sin(iz) = \frac{1}{2i}(e^{i(iz)} - e^{-i(iz)}) = -\tfrac{1}{2}i(e^{-z} - e^z) = i \sinh z,$$
and
$$\cos(iz) = \tfrac{1}{2}(e^{i(iz)} + e^{-i(iz)}) = \tfrac{1}{2}(e^{-z} + e^z) = \cosh z. \quad \blacksquare$$

The hyperbolic functions satisfy a number of basic identities, summarized in Theorem 4.5. We omit the proofs, which can all be deduced either from Theorem 4.3, by using the identities in Theorem 4.4, or directly from the definitions.

Theorem 4.5 Hyperbolic Identities

(a) **Addition**

$$\sinh(z_1 + z_2) = \sinh z_1 \cosh z_2 + \cosh z_1 \sinh z_2;$$
$$\cosh(z_1 + z_2) = \cosh z_1 \cosh z_2 + \sinh z_1 \sinh z_2;$$
$$\tanh(z_1 + z_2) = \frac{\tanh z_1 + \tanh z_2}{1 + \tanh z_1 \tanh z_2}.$$

(b) **Squares**

$$\cosh^2 z - \sinh^2 z = 1; \ \operatorname{sech}^2 z = 1 - \tanh^2 z; \ \operatorname{cosech}^2 z = \coth^2 z - 1.$$

(c) **Negatives**

$$\sinh(-z) = -\sinh z; \ \cosh(-z) = \cosh z; \ \tanh(-z) = -\tanh z.$$

(d) **Periodicity**

$$\sinh(z + 2\pi i) = \sinh z; \ \cosh(z + 2\pi i) = \cosh z; \ \tanh(z + \pi i) = \tanh z.$$

The following example and problem show that the hyperbolic functions play an important role in the determination of the real and imaginary parts of $\sin z$ and $\cos z$.

Example 4.4

Let $z = x + iy$. Prove the following.

(a) $\sin z = \sin x \cosh y + i \cos x \sinh y$ (b) $|\sin z|^2 = \sin^2 x + \sinh^2 y$

Solution

(a) We have

$$\sin(x + iy) = \sin x \cos(iy) + \cos x \sin(iy) \quad \text{(by Theorem 4.3(a))}$$
$$= \sin x \cosh y + i \cos x \sinh y \quad \text{(by Theorem 4.4)}.$$

(b) We have

$$|\sin(x + iy)|^2 = \sin^2 x \cosh^2 y + \cos^2 x \sinh^2 y \quad \text{(by part (a))}$$
$$= \sin^2 x (1 + \sinh^2 y) + \cos^2 x \sinh^2 y$$
$$= \sin^2 x + \sinh^2 y (\sin^2 x + \cos^2 x)$$
$$= \sin^2 x + \sinh^2 y. \quad \blacksquare$$

Problem 4.7

Let $z = x + iy$. Prove the following.

(a) $\cos z = \cos x \cosh y - i \sin x \sinh y$ (b) $|\cos z|^2 = \cos^2 x + \sinh^2 y$

5 LOGARITHMS AND POWERS

After working through this section, you should be able to:

(a) determine the *principal logarithm*, Log z, of a non-zero complex number z, and describe the geometric effect of the function $z \longmapsto$ Log z;

(b) determine the *principal power*, z^α, where $\alpha \in \mathbb{C}$, of a non-zero complex number z.

5.1 Logarithms of complex numbers

In real analysis the natural logarithm function (that is, logarithm to base e) is defined as the inverse of the exponential function. Since $x \longmapsto e^x$ is a one-one function on $\mathbb{R}$ with image $]0, \infty[$, it has an inverse function, $\log_e$, with domain $]0, \infty[$, defined by the rule

$$\log_e y = x, \quad \text{where } y = e^x.$$

The graph of the function $\log_e$ may be obtained by reflecting the graph $y = e^x$ in the line $y = x$ (Figure 5.1).

Now consider the complex exponential function $f(z) = e^z$. In trying to define an inverse function f^{-1} for f, we encounter the fact that f is not a one-one function. For example,

$$f(0) = f(2\pi i) = f(4\pi i) = \ldots = 1.$$

To get around this difficulty we seek a set A (preferably as large as possible) on which $f(z) = e^z$ is one-one. Taking this set A as the domain of f, we then define an inverse function f^{-1} with domain $f(A)$. This approach was used in Subsection 1.5 with the function $f(z) = z^2$, and we again find that there are many choices for the set A. The choice in the following example is motivated by the periodicity property

$$e^{z+2\pi i} = e^z, \quad \text{for all } z \text{ in } \mathbb{C},$$

which implies that A cannot contain two points which differ by $2\pi i$.

Some texts write ln for $\log_e$.

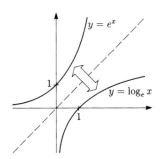

Figure 5.1

Thus we consider the restriction of f to the set A, and retain the name f for this restriction.

Example 5.1

Let

$$A = \{x + iy : -\pi < y \leq \pi\},$$

which is a horizontal strip (Figure 5.2). Prove that the function

$$f(z) = e^z \quad (z \in A)$$

has an inverse function f^{-1}, and determine the domain and rule of f^{-1}.

Solution

First we determine the image of f:

$$\begin{aligned} f(A) &= \{e^z : z \in A\} \\ &= \{w = e^{x+iy} : x \in \mathbb{R}, -\pi < y \leq \pi\} \\ &= \{w = e^x e^{iy} : x \in \mathbb{R}, -\pi < y \leq \pi\} \\ &= \{w = \rho e^{i\phi} : \rho > 0, -\pi < \phi \leq \pi\} \quad (\rho = e^x, \phi = y) \\ &= \mathbb{C} - \{0\}. \end{aligned}$$

Now, for each $w \in \mathbb{C} - \{0\}$, we wish to solve the equation

$$w = e^z \tag{5.1}$$

to obtain a unique solution z in A.

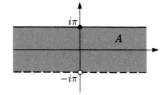

Figure 5.2

We have already seen that $f(A) = \mathbb{C} - \{0\}$ in Figure 4.7.

Each w in $\mathbb{C} - \{0\}$ can be written in the form
$$w = \rho e^{i\phi}, \quad \text{where } \rho > 0 \text{ and } -\pi < \phi \leq \pi,$$
and Equation (5.1) is then
$$\rho e^{i\phi} = e^z = e^x e^{iy}, \quad \text{where } z = x + iy.$$
Thus x and y must satisfy
$$\rho = e^x \quad \text{and} \quad e^{i\phi} = e^{iy};$$
that is,
$$x = \log_e \rho \quad \text{and} \quad y = \phi + 2n\pi, \quad \text{where } n \in \mathbb{Z}.$$
For $n = 0$, the solution is
$$z = x + iy = \log_e \rho + i\phi,$$
which lies in A, since $-\pi < \phi \leq \pi$, whereas the other solutions (with $n \neq 0$) lie outside A.

Thus f is a one-one function with image $\mathbb{C} - \{0\}$. Hence f has an inverse function f^{-1} with domain $\mathbb{C} - \{0\}$ and rule
$$f^{-1}(w) = \log_e \rho + i\phi, \quad w = \rho e^{i\phi}, \rho > 0, -\pi < \phi \leq \pi. \quad \blacksquare$$

Remark Since $\phi = \operatorname{Arg} w$ and $\rho = |w|$ in this solution, the rule for f^{-1} can be written in the form
$$f^{-1}(w) = \log_e |w| + i \operatorname{Arg} w \quad (w \neq 0).$$

Problem 5.1

Let
$$A = \{x + iy : 0 \leq y < 2\pi\}.$$
Prove that the function
$$f(z) = e^z \quad (z \in A)$$
has an inverse function f^{-1}, and determine the domain and rule of f^{-1}.

The solution to Example 5.1 and the Remark show that if $w \neq 0$, then the equation $e^z = w$ has infinitely many solutions of the form
$$z = \log_e |w| + i \, (\operatorname{Arg} w + 2n\pi), \quad n \in \mathbb{Z}.$$
Each of these solutions is called **a logarithm of w** (written $\log w$ if it is clear which particular logarithm is intended). These logarithms of w correspond to the infinitely many arguments of w, which are of the form
$$\operatorname{Arg} w + 2n\pi, \quad n \in \mathbb{Z}.$$
To avoid confusion, we shall almost always use the logarithm of w which corresponds to the principal argument of w (that is, $n = 0$); this solution,
$$z = \log_e |w| + i \operatorname{Arg} w,$$
of $e^z = w$ is called the *principal logarithm* of w, written $\operatorname{Log} w$. Thus the inverse function f^{-1} of Example 5.1 can be written as
$$f^{-1}(w) = \operatorname{Log} w \quad (w \in \mathbb{C} - \{0\}).$$

Note the capital L in Log, corresponding to the capital A in Arg.

> **Definition** For $z \in \mathbb{C} - \{0\}$, the **principal logarithm** of z is
> $$\operatorname{Log} z = \log_e |z| + i \operatorname{Arg} z.$$
> The corresponding **principal logarithm function** is called Log.

Remarks

1 Note that if z is real and positive (that is, $z = x + 0i$, where $x > 0$), then

$$\text{Log } z = \text{Log } x = \log_e x,$$

as expected. Thus, the restriction of Log to $]0, \infty[$ is $\log_e$.

2 Note that the definition applies if z is a negative real number. For example,

$$\text{Log}(-2) = \log_e |-2| + i \,\text{Arg}(-2) = \log_e 2 + i\pi.$$

3 Since Log is the inverse of the function

$$f(z) = e^z \qquad (z \in \{x + iy : -\pi < y \leq \pi\}),$$

we have two identities.

$$e^{\text{Log } z} = z, \quad \text{for } z \in \mathbb{C} - \{0\},$$

and

$$\text{Log}(e^z) = z, \quad \text{for } z \in \{x + iy : -\pi < y \leq \pi\}.$$

Note that the latter identity is false if z lies outside the strip

$$\{x + iy : -\pi < y \leq \pi\}.$$

For example, if $z = 2\pi i$, then

$$\text{Log}(e^{2\pi i}) = \text{Log}(1)$$
$$= \log_e 1 = 0 \neq 2\pi i.$$

4 The function Log has domain $\mathbb{C} - \{0\}$ and its image, written in terms of w, is $\{w : -\pi < \text{Im } w \leq \pi\}$, as shown in Figure 5.3.

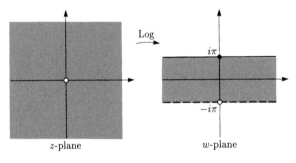

Figure 5.3

Example 5.2

Evaluate $\text{Log}(-1 + i)$ in Cartesian form.

Solution

Since $|-1 + i| = \sqrt{2}$ and $\text{Arg}(-1 + i) = 3\pi/4$, we have

$$\text{Log}(-1 + i) = \log_e \sqrt{2} + i\, 3\pi/4. \quad \blacksquare$$

Problem 5.2

Evaluate the following in Cartesian form.

(a) $\text{Log } i$ (b) $\text{Log}(\sqrt{3} - i)$ (c) $\text{Log}(\tfrac{1}{2} + \tfrac{1}{2}i)$

The real function $\log_e$ satisfies various identities, such as

$$\log_e(x_1 x_2) = \log_e x_1 + \log_e x_2, \quad \text{for } x_1, x_2 > 0,$$

and

$$\log_e(1/x) = -\log_e x, \quad \text{for } x > 0.$$

It is natural to hope that similar identities will hold for the complex function Log, and this is indeed the case, provided that suitable restrictions are placed on the variables involved.

> **Theorem 5.1 Logarithmic Identities**
> (a) **Multiplication**
> $\text{Log}(z_1 z_2) = \text{Log } z_1 + \text{Log } z_2, \quad \text{if Arg } z_1, \text{Arg } z_2 \in \,]-\tfrac{1}{2}\pi, \tfrac{1}{2}\pi].$
> (b) **Reciprocals**
> $\text{Log}(1/z) = -\text{Log } z, \quad \text{if Arg } z \in \,]-\pi, \pi[.$

As you can check using $z = -1$, part (b) does not hold if $\text{Arg } z = \pi$.

Proof
(a) If $\text{Arg } z_1, \text{Arg } z_2 \in \,]-\tfrac{1}{2}\pi, \tfrac{1}{2}\pi]$, then $\text{Arg } z_1 + \text{Arg } z_2 \in \,]-\pi, \pi]$, and so
$$\text{Arg}(z_1 z_2) = \text{Arg } z_1 + \text{Arg } z_2.$$

You met this property of Arg in *Unit A1*, Subsection 2.3.

Hence
$$\begin{aligned}\text{Log}(z_1 z_2) &= \log_e |z_1 z_2| + i \,\text{Arg}(z_1 z_2)\\ &= \log_e |z_1| + \log_e |z_2| + i\,(\text{Arg } z_1 + \text{Arg } z_2)\\ &= (\log_e |z_1| + i\,\text{Arg } z_1) + (\log_e |z_2| + i\,\text{Arg } z_2)\\ &= \text{Log } z_1 + \text{Log } z_2.\end{aligned}$$

(b) Since $1/z = \bar{z}/|z|^2$ and $\text{Arg } z \neq \pi$, it follows that
$$\text{Arg}(1/z) = \text{Arg } \bar{z} = -\text{Arg } z.$$

Hence
$$\begin{aligned}\text{Log}(1/z) &= \log_e |1/z| + i\,\text{Arg}(1/z)\\ &= -\log_e |z| - i\,\text{Arg } z\\ &= -\text{Log } z. \quad \blacksquare\end{aligned}$$

$|1/z| = 1/|z|$ and $\log_e 1 = 0$.

Remark The identity in part (a) of Theorem 5.1 holds in the following form for any values in the domain, $\mathbb{C} - \{0\}$, of Log:
$$\text{Log}(z_1 z_2) = \text{Log } z_1 + \text{Log } z_2 + 2n\pi i, \tag{5.2}$$

where n is -1, 0 or 1 according as $\text{Arg } z_1 + \text{Arg } z_2$ is greater than π, lies in the interval $]-\pi, \pi]$, or is less than or equal to $-\pi$.

For example, if $z_1 = z_2 = -1$, then
$$\text{Log}(z_1 z_2) = \text{Log } 1 = 0$$
and
$$\text{Log } z_1 + \text{Log } z_2 = i\pi + i\pi = 2\pi i.$$

Thus the identity (5.2) holds with $n = -1$ in this case.

The geometric nature of the function Log

We now discuss briefly the geometric effect of the function Log, drawing on the knowledge of the geometric effect of the exponential function $z \longmapsto e^z$ obtained in Subsection 4.1, and which is represented in Figure 4.6. In that figure the horizontal strip $\{z : -\pi < \text{Im } z \leq \pi\}$ is the domain of the restriction of $z \longmapsto e^z$ whose inverse function is Log. Thus Log maps circles with centre 0 and radius r onto line segments of the form $u = \log_e r$, $-\pi < v \leq \pi$ and it maps rays $\text{Arg } z = \theta$ onto lines of the form $v = \theta$.

$w = u + iv$

Figure 5.4 shows the image of this polar grid under Log.

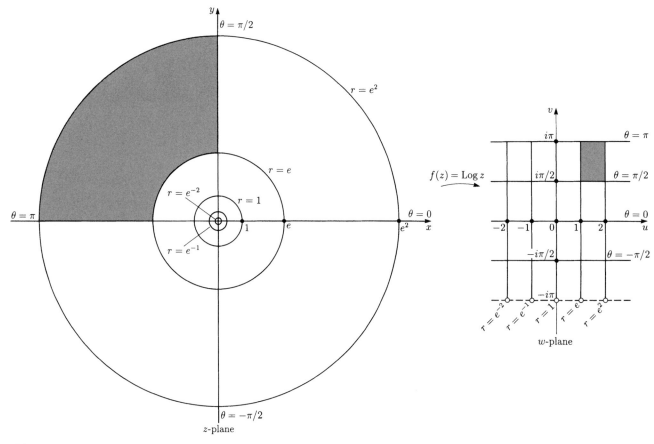

Figure 5.4

Notice that

(a) points lying outside the unit circle $\{z : |z| = 1\}$ have images lying in the right half-strip, whereas those non-zero points inside the unit circle have images lying in the left half-strip;

(b) the Log function behaves in a rather strange manner near the negative real axis — for example, as z approaches the point -1 on the negative real axis from *above*, the image point $w = \text{Log } z$ approaches the point $i\pi$, but if z approaches -1 from *below*, then $w = \text{Log } z$ approaches the point $-i\pi$ (which is not in the image of Log, of course). We consider this behaviour further in *Unit A3*.

This strange behaviour occurs near the negative real axis because of the particular definition of Arg which we have chosen. (Recall that
$$\text{Log } z = \log_e |z| + i \text{ Arg } z.)$$

Problem 5.3

Classify the following statements as True or False.

(a) The image under the function Log of the ellipse
$$4x^2 + 9y^2 = 1$$
lies in the right half-plane.

(b) The image under the function Log of the ray $\theta = \pi/4$ lies in the right half-plane.

(c) There is a point $z \in \mathbb{C}$ such that
$$\text{Log } z = 1 + 4i.$$

(d) There is a point $z \in \mathbb{C}$ such that
$$\text{Log } z = 1 + \tfrac{1}{4}i.$$

5.2 Powers of complex numbers

In this subsection we define the expression z^α, where z is any *non-zero* complex number and α is any complex number. In *Unit A1*, Subsection 3.1, you saw that any non-zero complex number z has n nth roots and that the expression $z^{1/n}$ is reserved for just one of these roots, called the *principal nth root of z*. It seems likely, therefore, that there is going to be some difficulty in defining the expression z^α in a unique way.

Recall first that if $a > 0$ and $x \in \mathbb{R}$, then a^x satisfies the equation

$$a^x = e^{x \log_e a}.$$

It is very tempting to define z^α by means of a similar formula, namely

$$z^\alpha = e^{\alpha \log z}, \quad \text{where } \log z \text{ is a logarithm of } z.$$

The problem is, however, that any non-zero complex number z has infinitely many logarithms, and so the above formula would give rise to infinitely many possible values of z^α. To avoid confusion, we shall define z^α using the principal logarithm of z, Log z.

Some texts allow both $\log z$ and z^α to represent infinitely many different values and specify, when appropriate, which value is being considered at a given time.

Definition For $z, \alpha \in \mathbb{C}$, with $z \neq 0$, the **principal αth power of z** is

$$z^\alpha = \exp(\alpha \operatorname{Log} z).$$

The function $z \longmapsto z^\alpha$ is called the **principal αth power function**.

Remarks

1 It can be shown that this definition agrees with the usual meaning of z^α if $\alpha = n$ or $\alpha = 1/n$, where n is a positive integer. For example, if $\alpha = n$ and $z \neq 0$, then

$$\begin{aligned} e^{n \operatorname{Log} z} &= e^{\operatorname{Log} z + \cdots + \operatorname{Log} z} \\ &= e^{\operatorname{Log} z} \times \cdots \times e^{\operatorname{Log} z} \\ &= z \times \cdots \times z = z^n. \end{aligned}$$

2 This definition assigns no value to 0^α. However, in *Unit A1*, we defined 0^n and $0^{1/n}$ to be 0, for $n = 1, 2, 3, \ldots$.

Example 5.3

Express each of the following numbers in Cartesian form.

(a) $(-1)^{1/2}$ (b) $(1+i)^i$ (c) i^i

Solution

(a) $\begin{aligned}(-1)^{1/2} &= \exp\left(\tfrac{1}{2} \operatorname{Log}(-1)\right) \\ &= e^{i\pi/2}, \quad \text{since } \operatorname{Log}(-1) = i\pi, \\ &= i \end{aligned}$

(b) $\begin{aligned}(1+i)^i &= \exp(i \operatorname{Log}(1+i)) \\ &= \exp(i(\log_e \sqrt{2} + i\pi/4)) \\ &= \exp(-\pi/4 + i \log_e \sqrt{2}) \\ &= e^{-\pi/4}(\cos(\log_e \sqrt{2}) + i \sin(\log_e \sqrt{2}))\end{aligned}$

$1 + i = \sqrt{2}\,(\cos \pi/4 + i \sin \pi/4)$

(c) $\begin{aligned} i^i &= \exp(i \operatorname{Log} i) \\ &= \exp(i(i\pi/2)), \quad \text{since } \operatorname{Log} i = i\pi/2, \\ &= e^{-\pi/2} \quad \text{(a real number)} \quad \blacksquare \end{aligned}$

Problem 5.4

Express each of the following numbers in Cartesian form.

(a) $(1+i)^{2/3}$ (b) i^{1+i}

Problem 5.5

Show that for $\alpha = 1/n$, where n is a positive integer, the definition of z^α given above agrees with the definition of $z^{1/n}$ given in *Unit A1*, Subsection 3.1.

Problem 5.6

(a) Determine which one of the following equations is *not* an identity.

   (i) $z^\alpha z^\beta = z^{\alpha+\beta}$ (ii) $z_1^\alpha z_2^\alpha = (z_1 z_2)^\alpha$

(b) Prove that the other equation is an identity.

Recall the convention given before Theorem 4.1.

EXERCISES

Section 1

Exercise 1.1 Write down the domain of each of the following functions.

(a) $f(z) = (z-1)^2$ (b) $f(z) = \dfrac{1}{z-1}$ (c) $f(z) = \dfrac{z}{z^2+1}$

(d) $f(z) = \dfrac{1}{\operatorname{Re} z}$ (e) $f(z) = \dfrac{1}{|z|-1}$ (f) $f(z) = \dfrac{1}{z^3+1}$

Exercise 1.2 Determine the image of each of the following functions.

(a) $f(z) = 2z+1$ (b) $f(z) = \dfrac{1}{z-1}$ (c) $f(z) = \dfrac{z}{z-1}$

(d) $f(z) = |z-1|$ (e) $f(z) = \operatorname{Re}(z+i)$ (f) $f(z) = |\operatorname{Arg} z|$

Exercise 1.3 Let $f(z) = \dfrac{z-1}{z}$ and $g(z) = \dfrac{z}{z-1}$. Determine the domain and rule of each of the following functions.

(a) $f+g$ (b) $3f - 2ig$ (c) fg (d) f/g

Exercise 1.4 For the functions f and g of Exercise 1.3 write down the domain and rule of each of the following functions.

(a) $f \circ g$ (b) $g \circ f$ (c) $f \circ f$

Exercise 1.5 Determine whether or not each of the functions f in Exercise 1.2 is one-one, and write down the inverse function of f, where possible.

Exercise 1.6 Let

$$A = \{0\} \cup \{z : -\pi/3 < \operatorname{Arg} z \le \pi/3\}.$$

Prove that the function

$$f(z) = z^3 \quad (z \in A)$$

has an inverse function f^{-1}, and determine the domain and rule of f^{-1}.

Section 2

Exercise 2.1 Determine the real and imaginary parts, $\operatorname{Re} f$ and $\operatorname{Im} f$, of each of the following functions.

(a) $f(z) = \bar{z}$ (b) $f(z) = iz$ (c) $f(z) = z^3$ (d) $f(z) = |z|$

Exercise 2.2 Sketch each of the following surfaces.
(a) $s = \text{Re}(\bar{z} + 1)$ (b) $s = \text{Im}(|z| + i)$

Exercise 2.3 Sketch each of the paths Γ with the following parametrizations.
(a) $\gamma(t) = 1 - it$ $(t \in \mathbb{R})$
(b) $\gamma(t) = i + (1 - i)t$ $(t \in [0, 1])$
(c) $\gamma(t) = \cos t - i \sin t$ $(t \in [0, 2\pi])$

Exercise 2.4 For each of the following parametrizations γ, find the equation of the corresponding path Γ in terms of x and y only. Sketch and classify the path in each case.
(a) $\gamma(t) = (1 - t)(1 + i) + ti$ $(t \in \mathbb{R})$
(b) $\gamma(t) = 2\cos t + 3i \sin t$ $(t \in [0, 2\pi])$
(c) $\gamma(t) = 1 + 2\cos t - (1 - 2\sin t)i$ $(t \in [0, 2\pi])$

Exercise 2.5 Determine the standard parametrization for the following.
(a) The circle with centre $1 - i$ and radius 3.
(b) The ellipse $2x^2 + 3y^2 = 6$.
(c) The parabola $8y^2 = x$.

Exercise 2.6 Sketch the path with parametrization
$$\gamma(t) = \tfrac{1}{2}(\cos t + i \sin t) - \tfrac{1}{4}(\cos 2t + i \sin 2t) \quad (t \in [-\pi, \pi])$$
by first plotting $\gamma(t)$ for $t = 0, \pm\tfrac{1}{4}\pi, \pm\tfrac{1}{2}\pi, \pm\tfrac{3}{4}\pi, \pm\pi$.

Verify that the equation of the path is
$$4(x^2 + y^2)^2 - \tfrac{3}{2}(x^2 + y^2) + \tfrac{1}{2}x = \tfrac{3}{64}.$$

Exercise 2.7 Determine the image under the function $f(z) = \sqrt{z}$ of each of the following sets.
(a) The negative x-axis. (b) The circle $|z| = 1$.

Section 3

Exercise 3.1 Sketch the images under each of the following functions f of the Cartesian grid and the polar grid shown in the following figures.

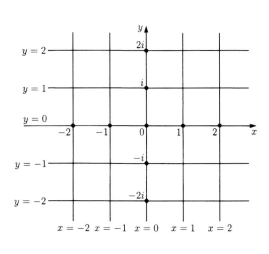
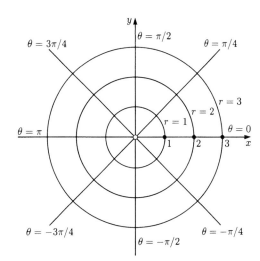

(a) $f(z) = z + i$ (b) $f(z) = 2z$ (c) $f(z) = 2 - iz$ (d) $f(z) = iz^2$

Section 4

Exercise 4.1 Express each of the following complex numbers in Cartesian form.
(a) $e^{3\pi i}$ (b) $ee^{\pi i/2}$ (c) $e^{2\pi i/3}$ (d) $e^{-3\pi i/2}$ (e) $e^{2+\pi i}$
(f) $e^{3+\pi i/2}$ (g) $e^{(\pi i/6)-1}$ (h) $e^{(\cos\theta + i\sin\theta)}$

Exercise 4.2
(a) Express each of the following complex numbers in the polar form $re^{i\theta}$.

  (i) $\dfrac{1}{\sqrt{2}} - \dfrac{i}{\sqrt{2}}$ (ii) $-(1+i)$ (iii) $1 + \sqrt{3}i$

(b) Hence evaluate

  (i) $\left(\dfrac{1}{\sqrt{2}} - \dfrac{i}{\sqrt{2}}\right)^3$ and   (ii) $(1 + \sqrt{3}i)^{-7}$,

giving your answers in Cartesian form.

Exercise 4.3 Express each of the following complex numbers in Cartesian form.
(a) $\sin(\pi + 2i)$ (b) $\cos(\pi/2 - i)$ (c) $\tan i$

In parts (a) and (b), you may work from the definitions of sin and cos, or use identities established in the section.

Exercise 4.4 Prove each of the following identities.
(a) $\overline{e^z} = e^{\bar{z}}$ (b) $\sin 2z = 2\sin z \cos z$ (c) $\overline{\sin z} = \sin \bar{z}$
(d) $\cosh(z_1 + z_2) = \cosh z_1 \cosh z_2 + \sinh z_1 \sinh z_2$
(e) $\cosh^2 z - \sinh^2 z = 1$

In parts (a), (b) and (c), work from the definitions of e^z, $\sin 2z$ and $\sin z$; in parts (d) and (e), use identities established in the section.

Exercise 4.5 Sketch the image under the function $f(z) = \sin z$ of the Cartesian grid shown in the following figure.

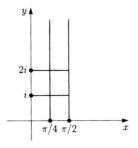

Section 5

Exercise 5.1 Express each of the following complex numbers in Cartesian form.
(a) $\text{Log}(-2)$ (b) $\text{Log}(i^3)$ (c) $\text{Log}(1+i)$ (d) $\text{Log}\sqrt{3}$
(e) $\text{Log}(i - \sqrt{3})$ (f) $\text{Log}\left(\dfrac{1-i}{\sqrt{2}}\right)$

Exercise 5.2 Express each of the following complex numbers in Cartesian form.
(a) i^{-i} (b) $(-i)^i$ (c) $(1-i)^i$ (d) $(1+i)^{1+i}$ (e) $(-1)^i$
(f) $(-1)^2$

SOLUTIONS TO THE PROBLEMS

Section 1

1.1 (a) Domain $\mathbb{C}$, codomain $\mathbb{C}$.
(b) Domain $\mathbb{C} - \{-2\}$, codomain $\mathbb{C}$.
(c) Domain $\mathbb{C} - \{0\}$, codomain $\mathbb{C}$.
(d) Domain $\mathbb{C} - \{-i, i\}$, codomain $\mathbb{C}$.

1.2 (a) The domain of f is $\mathbb{C}$. The image of f is
$$f(\mathbb{C}) = \{3iz : z \in \mathbb{C}\}$$
$$= \left\{w : z = \frac{w}{3i} \in \mathbb{C}\right\}$$
$$= \{w : w \in \mathbb{C}\}$$
$$= \mathbb{C}.$$

(b) The domain of f is $\mathbb{C} - \{-i\}$. The image of f is
$$f(\mathbb{C} - \{-i\}) = \left\{\frac{3z+1}{z+i} : z \in \mathbb{C} - \{-i\}\right\}$$
$$= \left\{w : z = \frac{1-iw}{w-3} \neq -i\right\}$$
$$= \{w : w \neq 3\}$$
$$\left(\frac{1-iw}{w-3} = -i \text{ has no solutions}\right)$$
$$= \mathbb{C} - \{3\}.$$

(c) The domain of f is $\mathbb{C}$. The image of f is
$$f(\mathbb{C}) = \{\operatorname{Im} z : z \in \mathbb{C}\}$$
$$= \{y : y \in \mathbb{R}\} \quad (z = x + iy)$$
$$= \mathbb{R}.$$

1.3 (a) f has domain $\mathbb{C}$ and the image of f is
$$f(\mathbb{C}) = \{x \in \mathbb{R} : x \geq 0\}.$$
(b) f has domain $\mathbb{C} - \{0\}$ and the image of f is
$$f(\mathbb{C} - \{0\}) = \;]-\pi, \pi].$$

1.4 (a) $f + g$ has domain $\mathbb{C} - \{0, 1\}$ and rule
$$(f+g)(z) = f(z) + g(z)$$
$$= \frac{1}{z} + \frac{z+3i}{z^2 - z}$$
$$= \frac{2z - 1 + 3i}{z^2 - z}.$$

(b) fg has domain $\mathbb{C} - \{0, 1\}$ and rule
$$(fg)(z) = f(z)g(z)$$
$$= \frac{1}{z} \cdot \frac{z+3i}{z^2 - z}$$
$$= \frac{z+3i}{z^3 - z^2}.$$

(c) f/g has domain $\mathbb{C} - \{0, 1, -3i\}$ and rule
$$(f/g)(z) = \frac{f(z)}{g(z)}$$
$$= \frac{1}{z} \Big/ \left(\frac{z+3i}{z^2 - z}\right)$$
$$= \frac{z-1}{z+3i}.$$
(Note that 0 and 1 are excluded from the domain of f/g even though $(z-1)/(z+3i)$ is defined at these points.)

1.5 (a) The domain of $g \circ f$ is
$$\text{domain of } f - \left\{z : \frac{1}{z} \notin \mathbb{C} - \{0, 1\}\right\}$$
$$= (\mathbb{C} - \{0\}) - \left\{z : \frac{1}{z} \in \{0, 1\}\right\}$$
$$= (\mathbb{C} - \{0\}) - \{1\} = \mathbb{C} - \{0, 1\}.$$
The rule of $g \circ f$ is
$$g(f(z)) = \frac{(1/z) + 3i}{(1/z)^2 - (1/z)} = \frac{z + 3iz^2}{1 - z}.$$

(b) The domain of $f \circ g$ is
$$\text{domain of } g - \left\{z : \frac{z+3i}{z^2 - z} \notin \mathbb{C} - \{0\}\right\}$$
$$= (\mathbb{C} - \{0, 1\}) - \left\{z : \frac{z+3i}{z^2 - z} \in \{0\}\right\}$$
$$= (\mathbb{C} - \{0, 1\}) - \{-3i\} = \mathbb{C} - \{0, 1, -3i\}.$$
The rule of $f \circ g$ is
$$f(g(z)) = 1 \Big/ \left(\frac{z+3i}{z^2 - z}\right)$$
$$= \frac{z^2 - z}{z + 3i}.$$

1.6 First we determine the image of f. This is $\mathbb{C} - \{3\}$, from Problem 1.2(b).
Now, for each $w \in \mathbb{C} - \{3\}$ we wish to solve the equation
$$w = \frac{3z+1}{z+i}$$
to obtain a unique solution z in $\mathbb{C} - \{-i\}$. This is achieved by the rearrangement
$$z = \frac{1 - iw}{w - 3}.$$
Thus f is a one-one function with image $\mathbb{C} - \{3\}$. Hence f has an inverse function f^{-1} with domain $\mathbb{C} - \{3\}$ and rule
$$f^{-1}(w) = \frac{1 - iw}{w - 3} \quad (w \in \mathbb{C} - \{3\}).$$

1.7 First we determine the image of f:
$$f(A) = \{w = z^2 : z \in A\}$$
$$= \{0\} \cup \{w = z^2 : 0 \leq \operatorname{Arg} z < \pi\} \quad (0^2 = 0)$$
$$= \{0\} \cup \{w = r^2(\cos 2\theta + i \sin 2\theta) : r > 0, 0 \leq \theta < \pi\}$$
$$\quad (z = r(\cos\theta + i\sin\theta))$$
$$= \{0\} \cup \{w = \rho(\cos\phi + i\sin\phi) : \rho > 0, 0 \leq \phi < 2\pi\}$$
$$\quad (\rho = r^2, \phi = 2\theta)$$
$$= \mathbb{C}.$$
Now, for each $w \in \mathbb{C}$ we wish to solve the equation
$$w = z^2 \tag{1}$$
to obtain a unique solution z in A. If $w = 0$, then Equation (1) has the unique solution $z = 0$. On the other hand, if $w \neq 0$, then w can be written in the form
$$w = \rho(\cos\phi + i\sin\phi), \quad \text{where } \rho > 0 \text{ and } 0 \leq \phi < 2\pi,$$
and Equation (1) then has exactly two solutions
$$z_0 = \rho^{1/2}\left(\cos\tfrac{1}{2}\phi + i\sin\tfrac{1}{2}\phi\right)$$
and
$$z_1 = \rho^{1/2}\left(\cos\left(\tfrac{1}{2}\phi + \pi\right) + i\sin\left(\tfrac{1}{2}\phi + \pi\right)\right),$$
by Theorem 3.1 of *Unit A1*. Clearly $z_0 \in A$, since $0 \leq \tfrac{1}{2}\phi < \pi$, whereas $z_1 \notin A$.

Thus f is a one-one function with image $\mathbb{C}$. Hence f has an inverse function f^{-1} with domain $\mathbb{C}$ and rule

$$f^{-1}(w) = \begin{cases} \rho^{1/2}\left(\cos\tfrac{1}{2}\phi + i\sin\tfrac{1}{2}\phi\right), & w = \rho(\cos\phi + i\sin\phi), \\ & \rho > 0, 0 \le \phi < 2\pi, \\ 0, & w = 0. \end{cases}$$

Remark Notice that the rule for f^{-1} is of the form
$$f^{-1}(w) = \sqrt{w}$$
for only some values of w (those for which $0 \le \phi < \pi$).

Section 2

2.1 (a)

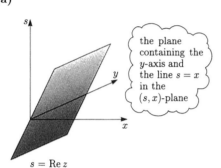

$s = \operatorname{Re} z$

(b)

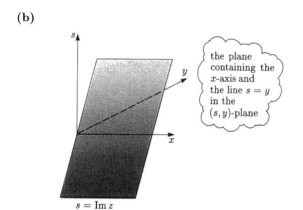

$s = \operatorname{Im} z$

2.2 $f(z) = \dfrac{1}{z}$ $(z \in \mathbb{C} - \{0\})$

$$= \frac{1}{x+iy}$$
$$= \frac{x-iy}{x^2+y^2}.$$

So
$$\operatorname{Re} f : z \longmapsto \frac{x}{x^2+y^2} \quad (z \in \mathbb{C} - \{0\}),$$
and
$$\operatorname{Im} f : z \longmapsto \frac{-y}{x^2+y^2} \quad (z \in \mathbb{C} - \{0\}).$$

2.3 (a) Since $\gamma(t) = 1 + it$ $(t \in \mathbb{R})$,
$$x = 1, \quad y = t.$$
Hence Γ is the line with equation $x = 1$, as shown.

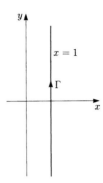

(b) Since $\gamma(t) = t^2 + it$ $(t \in [-1, 1])$,
$$x = t^2, \quad y = t. \qquad (1)$$
A brief table of values is as follows.

t	-1	$-\tfrac{1}{2}$	0	$\tfrac{1}{2}$	1
x	1	$\tfrac{1}{4}$	0	$\tfrac{1}{4}$	1
y	-1	$-\tfrac{1}{2}$	0	$\tfrac{1}{2}$	1

Hence the path Γ is as shown.

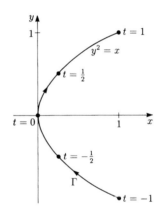

Eliminating t from Equations (1), we obtain
$$y^2 = x,$$
the equation of a parabola.

(c) Since $\gamma(t) = 1 - t + it$ $(t \in [0, 1])$,
$$x = 1 - t, \quad y = t. \qquad (2)$$
Eliminating t from Equations (2), we obtain
$$y = 1 - x,$$
the equation of a line.

When $t = 0$, we have $x = 1$, $y = 0$; when $t = 1$, we have $x = 0$, $y = 1$. Hence the path Γ is as shown.

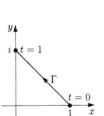

(d) Since $\gamma(t) = 2\cos t + 5i\sin t$ $(t \in [0, 2\pi])$,
$$x = 2\cos t, \quad y = 5\sin t. \qquad (3)$$
A brief table of values is as follows.

t	0	$\frac{1}{4}\pi$	$\frac{1}{2}\pi$	$\frac{3}{4}\pi$	π	$\frac{5}{4}\pi$	$\frac{3}{2}\pi$	$\frac{7}{4}\pi$	2π
x	2	$\sqrt{2}$	0	$-\sqrt{2}$	-2	$-\sqrt{2}$	0	$\sqrt{2}$	2
y	0	$\frac{5}{\sqrt{2}}$	5	$\frac{5}{\sqrt{2}}$	0	$\frac{-5}{\sqrt{2}}$	-5	$\frac{-5}{\sqrt{2}}$	0

Hence the path Γ is as shown.

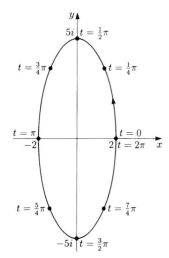

Eliminating t from Equations (3), we obtain
$$\frac{x^2}{4} + \frac{y^2}{25} = 1,$$
the equation of an ellipse.

2.4 In each case we use the table of standard parametrizations.

(a) $\gamma(t) = (1-t)(-2) + ti$
$= 2(t-1) + it \quad (t \in \mathbb{R})$.
Hence
$x = 2(t-1), \quad y = t$.

(b) $\gamma(t) = (1-t)(1) + t(1+i)$
$= 1 + ti \quad (t \in [0,1])$.
Hence
$x = 1, \quad y = t$.

(c) $\gamma(t) = (1+i) + 1(\cos t + i \sin t)$
$= 1 + \cos t + (1 + \sin t)i \quad (t \in [0, 2\pi])$.
Hence
$x = 1 + \cos t, \quad y = 1 + \sin t$.

(d) From the table, on interchanging the roles of x and y, we see that
$\gamma(t) = 2at + iat^2 \quad (t \in \mathbb{R})$
is the standard parametrization of the parabola
$x^2 = 4ay$.
Thus, for the parabola $y = x^2$, $a = \frac{1}{4}$ and the standard parametrization is
$\gamma(t) = \frac{1}{2}t + \frac{1}{4}it^2$.
Hence
$x = \frac{1}{2}t, \quad y = \frac{1}{4}t^2$.
(Of course, you may 'feel' that the parametrization
$\gamma(t) = t + it^2 \quad (t \in \mathbb{R})$
is simpler!)

Section 3

3.1

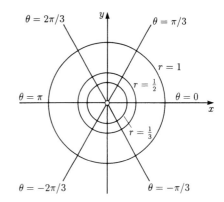

3.2 (a) $u = a - t, \quad v = a + t$;
adding these equations, we obtain
$u + v = 2a$.

(b) $u = a^2 - t^2, \quad v = 2at$;
hence
$$t^2 = \left(\frac{v}{2a}\right)^2,$$
and so
$$u = a^2 - \frac{v^2}{4a^2};$$
that is,
$v^2 = 4a^2(a^2 - u)$.

(c) $u = \dfrac{a}{a^2 + t^2}, \quad v = \dfrac{-t}{a^2 + t^2}$;
squaring each of these expressions and adding the results, we obtain
$$u^2 = \frac{a^2}{(a^2+t^2)^2}, \quad v^2 = \frac{t^2}{(a^2+t^2)^2}$$
and so
$$u^2 + v^2 = \frac{1}{a^2+t^2} = \frac{u}{a}.$$

3.3 The line $y = b$ has parametric equations
$x = t, \quad y = b \quad (t \in \mathbb{R})$.
Substituting these in
$u = x - y \quad \text{and} \quad v = x + y \quad \text{(from Frame 1)}$
gives the parametric equations of the image of the line $y = b$ under the function $f(z) = (1+i)z$. Thus
$u = t - b, \quad v = t + b$;
eliminating t, we obtain
$v - u = 2b$,
which is the equation of a line.
The images of the lines $y = 1$ and $y = 0$ are, respectively,
$v - u = 2 \quad \text{and} \quad v - u = 0$.
They are shown below, as are the directions of increasing t.

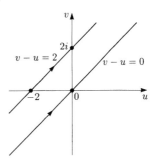

3.4 The line $y = b$ has parametric equations
$$x = t, \quad y = b \quad (t \in \mathbb{R}).$$
Substituting these in
$$u = x^2 - y^2 \quad \text{and} \quad v = 2xy \quad \text{(from Frame 5)}$$
gives the parametric equations of the image of the line $y = b$ under the function $f(z) = z^2$. Thus
$$u = t^2 - b^2, \quad v = 2tb;$$
eliminating t, we obtain
$$v^2 = 4b^2(u + b^2), \quad b \neq 0,$$
which is the equation of a parabola.

The images of the lines $y = 1$ and $y = 0$ are, respectively,
the parabola $v^2 = 4(u + 1)$
and
the non-negative u-axis $v = 0, u \geq 0$.

They are shown below, as are the directions of increasing t.

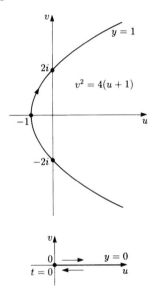

3.5 The line $y = b$ has parametric equations
$$x = t, \quad y = b \quad (t \in \mathbb{R}).$$
Substituting these in
$$u = \frac{x}{x^2 + y^2} \quad \text{and} \quad v = \frac{-y}{x^2 + y^2} \quad \text{(from Frame 9)}$$
gives the parametric equations of the image of the line $y = b$ under the function $f(z) = 1/z$. Thus
$$u = \frac{t}{t^2 + b^2}, \quad v = \frac{-b}{t^2 + b^2};$$
eliminating t, we obtain
$$u^2 + v^2 = -\frac{v}{b}, \quad b \neq 0,$$
which is the equation of a circle.

The images of the lines $y = 1$ and $y = 0$ are, respectively,
the circle $u^2 + v^2 + v = 0$, excluding the origin,
and
the line $v = 0, u \neq 0$.

They are shown at the top of the next column, as are the directions of increasing t.

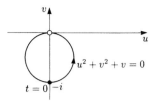

3.6 (a) The function $f(z) = iz + 1$ has the following geometric effect: it rotates the point z anticlockwise about the origin through $\pi/2$ and then translates the result to the right by one unit. Thus

the image of the line $y = 0$ is the line $u = 1$;

the image of the line $x = 1$ is the line $v = 1$;

the image of the line $y = 1$ is the line $u = 0$;

the image of the line $x = 0$ is the line $v = 0$;

and the image of S is S (in the w-plane), as shown in the figure.

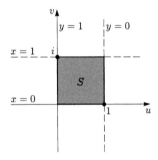

(b) Using the geometrical interpretation above, we obtain the image of the polar grid as follows.

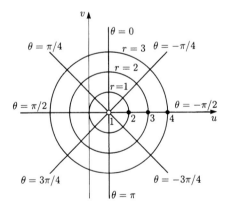

3.7 (a) Here we follow the approach using parametrization, and find the images of the lines $x = 0$, $x = 1$, $y = 0$ and $y = 1$.

The image of z under the function $f(z) = z^3$ is
$$\begin{aligned} w = f(z) &= z^3 \\ &= (x + iy)^3 \\ &= x^3 + 3x^2yi - 3xy^2 - iy^3 \\ &= (x^3 - 3xy^2) + (3x^2y - y^3)i, \end{aligned}$$
so that
$$u = x^3 - 3xy^2 \quad \text{and} \quad v = 3x^2y - y^3. \tag{1}$$

Line $x = 0$: This line has parametric equations
$$x = 0, \quad y = t \quad (t \in \mathbb{R});$$
substituting these in Equations (1), we obtain the parametric equations of the image
$$u = 0, \quad v = -t^3 \quad (t \in \mathbb{R}), \tag{2}$$
which is the v-axis.

Line $x = 1$: This line has parametric equations
$$x = 1, \quad y = t \quad (t \in \mathbb{R});$$
substituting these in Equations (1), we obtain the parametric equations of the image
$$u = 1 - 3t^2, \quad v = 3t - t^3 \quad (t \in \mathbb{R}). \tag{3}$$

Line $y = 0$: This line has parametric equations
$$x = t, \quad y = 0 \quad (t \in \mathbb{R});$$
substituting these in Equations (1), we obtain the parametric equations of the image
$$u = t^3, \quad v = 0 \quad (t \in \mathbb{R}), \tag{4}$$
which is the u-axis.

Line $y = 1$: This line has parametric equations
$$x = t, \quad y = 1 \quad (t \in \mathbb{R});$$
substituting these in Equations (1), we obtain the parametric equations of the image
$$u = t^3 - 3t, \quad v = 3t^2 - 1 \quad (t \in \mathbb{R}). \tag{5}$$

Restricting t to the interval $[0,1]$ in the parametric equations of the four lines, we obtain the boundary lines of the square
$$S = \{z : 0 \leq \operatorname{Re} z \leq 1, 0 \leq \operatorname{Im} z \leq 1\},$$
and Equations (2)–(5) restricted in the same way give their images. Thus

line $x = 0$ from 0 to i has image given by
$$u = 0, \quad v = -t^3, \quad (t \in [0,1]),$$
which is the v-axis from 0 to $-i$;

line $y = 0$ from 0 to 1 has image given by
$$u = t^3, \quad v = 0, \quad (t \in [0,1]),$$
which is the u-axis from 0 to 1.

We find the images of the other boundary lines by tabulating some values for t in the interval $[0,1]$.

Image of line $x = 1$

t	0	$\frac{1}{4}$	$1/\sqrt{3}$	$\frac{3}{4}$	1
u	1	0.81	0	-0.69	-2
v	0	0.73	1.54	1.83	2

Image of line $y = 1$

t	0	$\frac{1}{4}$	$1/\sqrt{3}$	$\frac{3}{4}$	1
u	0	-0.73	-1.54	-1.83	-2
v	-1	-0.81	0	0.69	2

Putting together this information, we obtain the image of S as shown below.

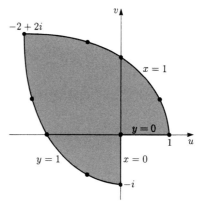

(b) If z has modulus r and argument θ, then $w = f(z) = z^3$ has modulus r^3 and argument 3θ. Thus the image of the polar grid (with the circle $r = 3$ omitted) is as shown in the figure.

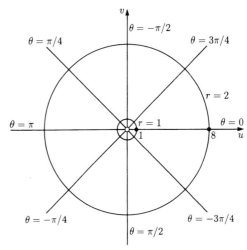

3.8 If z has modulus r and principal argument θ, then $w = f(z) = \sqrt{z}$ has modulus $r^{1/2}$ and principal argument $\theta/2$. Thus

the image of the ray $\theta = b$, where b is a constant in the interval $]-\pi, \pi]$, is the ray $\theta = b/2$;

the image of the circle with radius r and centre the origin is the semi-circle given by
$$|w| = \sqrt{r}, \quad \theta \in \,]-\pi/2, \pi/2].$$

Hence the image of the polar grid is shown below.

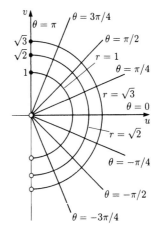

Section 4

4.1 (a) $e^{2\pi i} = e^0(\cos 2\pi + i \sin 2\pi) = 1$

(b) $e^{2+i\pi/3} = e^2\left(\cos\dfrac{\pi}{3} + i\sin\dfrac{\pi}{3}\right) = \dfrac{e^2(1+\sqrt{3}i)}{2}$

(c) $e^{-(1+i\pi)} = e^{-1}(\cos(-\pi) + i\sin(-\pi)) = -1/e$

4.2 (a) $e^{z+2\pi i} = e^z e^{2\pi i}$ (Theorem 4.1(a))
$= e^z$ ($e^{2\pi i} = 1$, see Problem 4.1(a)).

(b) $|e^z| = e^{\operatorname{Re} z}$ (Theorem 4.1(b))
$\le e^{|z|}$ ($|z| \ge \operatorname{Re} z$, and $x \mapsto e^x$ is an increasing function).

(c) The (complex) function exp is not one-one; this follows directly from part (a) above.

(d) (i) Since
$$e^{x+iy} = 1 \iff e^x = 1 \text{ and } \cos y + i\sin y = 1$$
$$\iff x = 0 \text{ and } y = 2n\pi, \text{ where } n \in \mathbb{Z},$$
it follows that
$$\{z : e^z = 1\} = \{2n\pi i : n \in \mathbb{Z}\}.$$

(ii) Since
$$e^{x+iy} = -1 \iff e^x = 1 \text{ and } \cos y + i\sin y = -1$$
$$\iff x = 0 \text{ and } y = (2n+1)\pi, \text{ where } n \in \mathbb{Z},$$
it follows that
$$\{z : e^z = -1\} = \{(2n+1)\pi i : n \in \mathbb{Z}\}.$$

4.3 Since $\sqrt{3} + i = 2\left(\cos\dfrac{\pi}{6} + i\sin\dfrac{\pi}{6}\right) = 2e^{i\pi/6}$,
$$(\sqrt{3}+i)^{-6} = (2e^{i\pi/6})^{-6}$$
$$= 2^{-6}(e^{i\pi/6})^{-6}$$
$$= 2^{-6}e^{-6 \times i\pi/6}$$
$$= 2^{-6}e^{-i\pi} = -1/64.$$

4.4 (a)

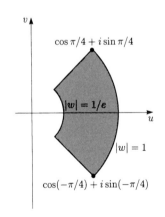

(b)

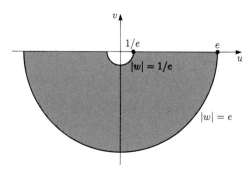

(c)

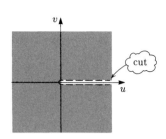

4.5 (a) $\sin(\pi/2 + i) = \dfrac{1}{2i}(e^{i(\pi/2+i)} - e^{-i(\pi/2+i)})$
$= \dfrac{1}{2i}(e^{-1+i\pi/2} - e^{1-i\pi/2})$
$= \dfrac{1}{2i}(e^{-1}i - e(-i))$
$= \dfrac{e^{-1}+e}{2}$

(b) $\cos i = \tfrac{1}{2}(e^{i\times i} + e^{-i \times i})$
$= \tfrac{1}{2}(e^{-1} + e)$ (which equals $\sin(\pi/2+i)$)

4.6 (a) (i) $\sin(-z) = \dfrac{1}{2i}(e^{i(-z)} - e^{-i(-z)})$
$= \dfrac{1}{2i}(e^{-iz} - e^{iz})$
$= -\dfrac{1}{2i}(e^{iz} - e^{-iz})$
$= -\sin z$

(ii) $\cos(z+2\pi) = \tfrac{1}{2}(e^{i(z+2\pi)} + e^{-i(z+2\pi)})$
$= \tfrac{1}{2}(e^{iz}e^{2\pi i} + e^{-iz}e^{-2\pi i})$
$= \tfrac{1}{2}(e^{iz} + e^{-iz})$
$= \cos z$

(b) (i) $\cos 2z = \cos(z+z)$
$= \cos z \cos z - \sin z \sin z$ (Theorem 4.3(a))
$= \cos^2 z - \sin^2 z$
$= 2\cos^2 z - 1$ (Theorem 4.3(b))

(ii) $\tan(z_1 - z_2) = \tan(z_1 + (-z_2))$
$= \dfrac{\tan z_1 + \tan(-z_2)}{1 - \tan z_1 \tan(-z_2)}$ (Theorem 4.3(a))
$= \dfrac{\tan z_1 - \tan z_2}{1 + \tan z_1 \tan z_2}$ (Theorem 4.3(c))

4.7 (a) $\cos z = \cos(x+iy)$
$= \cos x \cos(iy) - \sin x \sin(iy)$
(Theorem 4.3(a))
$= \cos x \cosh y - \sin x(i \sinh y)$
$= \cos x \cosh y - i \sin x \sinh y$

(b) $|\cos z|^2 = \cos^2 x \cosh^2 y + \sin^2 x \sinh^2 y$
(using part (a))
$= \cos^2 x(1 + \sinh^2 y) + \sin^2 x \sinh^2 y$
$= \cos^2 x + \sinh^2 y(\cos^2 x + \sin^2 x)$
$= \cos^2 x + \sinh^2 y$

Section 5

5.1 First we determine the image of f:
$$\begin{aligned}f(A) &= \{e^z : z \in A\} \\ &= \{w = e^{x+iy} : x \in \mathbb{R}, 0 \le y < 2\pi\} \\ &= \{w = e^x e^{iy} : x \in \mathbb{R}, 0 \le y < 2\pi\} \\ &= \{w = \rho e^{i\phi} : \rho > 0, 0 \le \phi < 2\pi\} \quad (\rho = e^x, \phi = y) \\ &= \mathbb{C} - \{0\}.\end{aligned}$$

Now, for each $w \in \mathbb{C} - \{0\}$, we wish to solve the equation
$$w = e^z \tag{1}$$
to obtain a unique solution z in A. Each w in $\mathbb{C} - \{0\}$ can be written in the form
$$w = \rho e^{i\phi}, \text{ where } \rho > 0 \text{ and } 0 \le \phi < 2\pi,$$
and Equation (1) is then
$$\rho e^{i\phi} = e^z = e^x e^{iy}, \quad \text{where } z = x + iy.$$
Thus, x and y must satisfy
$$\rho = e^x \quad \text{and} \quad e^{i\phi} = e^{iy};$$
that is
$$x = \log_e \rho \quad \text{and} \quad y = \phi + 2n\pi, \quad \text{where } n \in \mathbb{Z}.$$
For $n = 0$, the solution is
$$z = x + iy = \log_e \rho + i\phi,$$
which lies in A, since $0 \le \phi < 2\pi$, whereas the other solutions (with $n \ne 0$) lie outside A.

Thus f is a one-one function with image $\mathbb{C} - \{0\}$. Hence f has inverse function f^{-1} with domain $\mathbb{C} - \{0\}$ and rule
$$f^{-1}(w) = \log_e \rho + i\phi, \quad w = \rho e^{i\phi}, \rho > 0, 0 \le \phi < 2\pi.$$

5.2 (a) $\begin{aligned}[t]\operatorname{Log} i &= \log_e |i| + i \operatorname{Arg} i \\ &= \log_e 1 + i\frac{\pi}{2} = \frac{\pi}{2}i\end{aligned}$

(b) $\begin{aligned}[t]\operatorname{Log}(\sqrt{3} - i) &= \log_e |\sqrt{3} - i| + i \operatorname{Arg}(\sqrt{3} - i) \\ &= \log_e 2 - \frac{\pi}{6}i\end{aligned}$

(c) $\begin{aligned}[t]\operatorname{Log}\left(\tfrac{1}{2} + \tfrac{1}{2}i\right) &= \log_e \left|\tfrac{1}{2} + \tfrac{1}{2}i\right| + i \operatorname{Arg}\left(\tfrac{1}{2} + \tfrac{1}{2}i\right) \\ &= \log_e \frac{1}{\sqrt{2}} + i\frac{\pi}{4} \\ &= -\log_e \sqrt{2} + \frac{\pi}{4}i\end{aligned}$

5.3 (a) False. (The ellipse $4x^2 + 9y^2 = 1$ lies entirely inside the unit circle $|z| = 1$, and so its image lies in the left half-plane.)

(b) False. (The ray $\theta = \pi/4$ lies inside, on and outside the unit circle $|z| = 1$, and so its image is not confined to the right half-plane.)

(c) False. (Since $1 + 4i$ does not lie in the strip $\{w : -\pi < \operatorname{Im} w \le \pi\}$, which is the image of the function Log, there is no $z \in \mathbb{C}$ such that $\operatorname{Log} z = 1 + 4i$.)

(d) True. (Since $1 + \tfrac{1}{4}i \in \{w : -\pi < \operatorname{Im} w \le \pi\}$, there is a $z \in \mathbb{C}$ such that $\operatorname{Log} z = 1 + \tfrac{1}{4}i$.)

5.4 (a) $\begin{aligned}[t](1+i)^{2/3} &= \exp\left(\tfrac{2}{3} \operatorname{Log}(1+i)\right) \\ &= \exp\left(\tfrac{2}{3}\left(\log_e \sqrt{2} + i\pi/4\right)\right) \\ &= \exp\left(\tfrac{1}{3}\log_e 2 + i\pi/6\right) \\ &= 2^{1/3} e^{i\pi/6} \\ &= 2^{1/3}\left(\frac{\sqrt{3}}{2} + \frac{1}{2}i\right) = 2^{-2/3}(\sqrt{3} + i)\end{aligned}$

(b) $\begin{aligned}[t]i^{1+i} &= \exp((1+i)\operatorname{Log} i) \\ &= \exp\left((1+i)\frac{\pi}{2}i\right) \quad (\text{Problem 5.2(a)}) \\ &= e^{-\pi/2} e^{i\pi/2} \\ &= e^{-\pi/2} i\end{aligned}$

5.5 Since $z^\alpha = \exp(\alpha \operatorname{Log} z)$,
$$\begin{aligned}z^{1/n} &= \exp\left(\frac{1}{n}\operatorname{Log} z\right) \\ &= \exp\left(\frac{1}{n}(\log_e |z| + i\operatorname{Arg} z)\right) \\ &= \exp\left(\frac{1}{n}(\log_e \rho + i\phi)\right) \quad (\rho = |z|, \phi = \operatorname{Arg} z) \\ &= \exp\left(\log_e \rho^{1/n} + i\frac{\phi}{n}\right) \\ &= \rho^{1/n} e^{i\phi/n} \\ &= \rho^{1/n}\left(\cos\frac{\phi}{n} + i\sin\frac{\phi}{n}\right),\end{aligned}$$
which is the principal nth root of $z = \rho(\cos\phi + i\sin\phi)$ because ϕ is the principal argument of z (see *Unit A1*, Subsection 3.1).

5.6 (a) $z_1^\alpha z_2^\alpha = (z_1 z_2)^\alpha$ is not an identity. Consider $z_1 = -1 = z_2$, and $\alpha = \tfrac{1}{2}$. Then
$$\begin{aligned}z_1^\alpha z_2^\alpha &= (-1)^{1/2}(-1)^{1/2} \\ &= i \times i \quad (\text{Example 5.3(a)}) \\ &= -1.\end{aligned}$$
However
$$\begin{aligned}(z_1 z_2)^\alpha &= (-1 \times -1)^{1/2} \\ &= 1^{1/2} \\ &= 1 (\ne -1),\end{aligned}$$
so $z_1^\alpha z_2^\alpha = (z_1 z_2)^\alpha$ is not an identity.

(b) We have to prove the identity
$$z^\alpha z^\beta = z^{\alpha + \beta}.$$
By definition,
$$z^\alpha = \exp(\alpha(\log_e |z| + i\operatorname{Arg} z))$$
and
$$z^\beta = \exp(\beta(\log_e |z| + i\operatorname{Arg} z));$$
so
$$\begin{aligned}z^\alpha z^\beta &= e^{\alpha(\log_e |z| + i\operatorname{Arg} z)} e^{\beta(\log_e |z| + i\operatorname{Arg} z)} \\ &= e^{\alpha(\log_e |z| + i\operatorname{Arg} z) + \beta(\log_e |z| + i\operatorname{Arg} z)} \\ &\qquad (\text{by Theorem 4.1(a)}) \\ &= e^{(\alpha + \beta)(\log_e |z| + i\operatorname{Arg} z)} \\ &= \exp((\alpha + \beta)(\log_e |z| + i\operatorname{Arg} z)) \\ &= z^{\alpha + \beta} \quad (\text{by definition}),\end{aligned}$$
as required.

SOLUTIONS TO THE EXERCISES

Section 1

1.1 (a) $\mathbb{C}$ (b) $\mathbb{C} - \{1\}$ (c) $\mathbb{C} - \{-i, i\}$
(d) $\mathbb{C} - \{z : \operatorname{Re} z = 0\} = \{z : \operatorname{Re} z \neq 0\}$
(e) $\mathbb{C} - \{z : |z| = 1\} = \{z : |z| \neq 1\}$
(f) $\mathbb{C} - \left\{ \dfrac{1 + \sqrt{3}i}{2}, -1, \dfrac{1 - \sqrt{3}i}{2} \right\}$

1.2 The images of f are determined as follows.
(a) $\{2z + 1 : z \in \mathbb{C}\} = \{w : z = (w - 1)/2 \in \mathbb{C}\}$
$\qquad = \{w : w \in \mathbb{C}\} = \mathbb{C}$
(b) $\{1/(z - 1) : z \in \mathbb{C} - \{1\}\}$
$\qquad = \{w : z = (1 + w)/w \neq 1\}$
$\qquad = \{w : w \neq 0\} = \mathbb{C} - \{0\}$
(c) $\{z/(z - 1) : z \in \mathbb{C} - \{1\}\}$
$\qquad = \{w : z = w/(w - 1) \neq 1\}$
$\qquad = \{w : w \neq 1\} = \mathbb{C} - \{1\}$
(d) $\{|z - 1| : z \in \mathbb{C}\} = \{r : r \geq 0\}$
$\qquad = [0, \infty[$
(e) $\{\operatorname{Re}(z + i) : z \in \mathbb{C}\} = \{x : x \in \mathbb{R}\}$ $(z = x + iy)$
$\qquad = \mathbb{R}$
(f) $\{|\operatorname{Arg} z| : z \in \mathbb{C} - \{0\}\} = \{\theta : \theta \in [0, \pi]\}$
$\qquad = [0, \pi]$

1.3 (a) $f + g$ has domain $\mathbb{C} - \{0, 1\}$ and rule
$$(f + g)(z) = \frac{z - 1}{z} + \frac{z}{z - 1}$$
$$= \frac{2z^2 - 2z + 1}{z(z - 1)}.$$
(b) $3f$ has domain $\mathbb{C} - \{0\}$ and $2ig$ has domain $\mathbb{C} - \{1\}$; hence $3f - 2ig$ has domain $\mathbb{C} - \{0, 1\}$ and rule
$$(3f - 2ig)(z) = \frac{3(z - 1)}{z} - \frac{2iz}{z - 1}$$
$$= \frac{3z^2 - 6z + 3 - 2iz^2}{z(z - 1)}$$
$$= \frac{(3 - 2i)z^2 - 6z + 3}{z(z - 1)}.$$
(c) fg has domain $\mathbb{C} - \{0, 1\}$ and rule
$$(fg)(z) = \frac{z - 1}{z} \cdot \frac{z}{z - 1}$$
$$= 1.$$
(d) f/g has domain $(\mathbb{C} - \{0, 1\}) - \{0\} = \mathbb{C} - \{0, 1\}$ and rule
$$(f/g)(z) = \frac{z - 1}{z} \bigg/ \frac{z}{z - 1}$$
$$= \left(\frac{z - 1}{z} \right)^2.$$

1.4 (a) The domain of $f \circ g$ is
$$\text{domain of } g - \left\{ z : \frac{z}{z - 1} \notin \mathbb{C} - \{0\} \right\}$$
$$= (\mathbb{C} - \{1\}) - \{0\} = \mathbb{C} - \{0, 1\};$$
the rule of $f \circ g$ is
$$f(g(z)) = \frac{\dfrac{z}{z - 1} - 1}{\dfrac{z}{z - 1}}$$
$$= 1 - \frac{z - 1}{z}$$
$$= \frac{1}{z}.$$
(b) The domain of $g \circ f$ is
$$\text{domain of } f - \left\{ z : \frac{z - 1}{z} \notin \mathbb{C} - \{1\} \right\}$$
$$= (\mathbb{C} - \{0\}) - \left\{ z : \frac{z - 1}{z} = 1 \right\}$$
$$= (\mathbb{C} - \{0\}) - \varnothing = \mathbb{C} - \{0\};$$
the rule of $g \circ f$ is
$$g(f(z)) = \frac{\dfrac{z - 1}{z}}{\dfrac{z - 1}{z} - 1}$$
$$= \frac{z - 1}{z - 1 - z}$$
$$= 1 - z.$$
(c) The domain of $f \circ f$ is
$$\text{domain of } f - \left\{ z : \frac{z - 1}{z} \notin \mathbb{C} - \{0\} \right\}$$
$$= (\mathbb{C} - \{0\}) - \{z : z - 1 = 0\}$$
$$= \mathbb{C} - \{0, 1\};$$
the rule of $f \circ f$ is
$$f(f(z)) = \frac{\dfrac{z - 1}{z} - 1}{\dfrac{z - 1}{z}}$$
$$= 1 - \frac{z}{z - 1}$$
$$= \frac{1}{1 - z}.$$

1.5 The functions in parts (d), (e) and (f) are not one-one, because (for example)
(d) $\quad f(2) = |2 - 1| = 1$ and $f(0) = |0 - 1| = 1$;
(e) $\quad f(1 + i) = \operatorname{Re}(1 + i + i) = 1$
and
$\quad f(1 - i) = \operatorname{Re}(1 - i + i) = 1;$
(f) $\quad f(i) = |\operatorname{Arg} i| = |\pi/2| = \pi/2$
and
$\quad f(-i) = |\operatorname{Arg}(-i)| = |-\pi/2| = \pi/2.$

The functions in parts (a), (b) and (c) are one-one, as we now show. We already know their images from Exercise 1.2.

(a) For each $w \in \mathbb{C}$, the image of f, we wish to solve the equation
$$w = 2z + 1$$

54

to obtain a unique solution z in $\mathbb{C}$, the domain of f. This is achieved by the rearrangement
$$z = (w - 1)/2.$$
Thus f is a one-one function with image $\mathbb{C}$. Hence f has an inverse function f^{-1} with domain $\mathbb{C}$ and rule
$$f^{-1}(w) = (w - 1)/2 \quad (w \in \mathbb{C}).$$
(b) For each $w \in \mathbb{C} - \{0\}$, the image of f, we wish to solve the equation
$$w = 1/(z - 1)$$
to obtain a unique solution z in $\mathbb{C} - \{1\}$, the domain of f. This is achieved by the rearrangement
$$z = (1 + w)/w.$$
Thus f is a one-one function with image $\mathbb{C} - \{0\}$. Hence f has an inverse function f^{-1} with domain $\mathbb{C} - \{0\}$ and rule
$$f^{-1}(w) = (1 + w)/w \quad (w \in \mathbb{C} - \{0\}).$$
(c) For each $w \in \mathbb{C} - \{1\}$, the image of f, we wish to solve the equation
$$w = z/(z - 1)$$
to obtain a unique solution z in $\mathbb{C} - \{1\}$, the domain of f. This is achieved by the rearrangement
$$z = w/(w - 1).$$
Thus f is a one-one function with image $\mathbb{C} - \{1\}$. Hence f has an inverse function f^{-1} with domain $\mathbb{C} - \{1\}$ and rule
$$f^{-1}(w) = w/(w - 1) \quad (w \in \mathbb{C} - \{1\}).$$
(Note that $f^{-1} = f$ in this case.)

1.6 First we determine the image of f:
$$\begin{aligned}f(A) &= \{z^3 : z \in A\} \\ &= \{0\} \cup \{w = z^3 : -\pi/3 < \operatorname{Arg} z \le \pi/3\} \quad (0^3 = 0) \\ &= \{0\} \cup \{w = r^3(\cos 3\theta + i\sin 3\theta) : r > 0, -\pi/3 < \theta \le \pi/3\} \\ &\quad (z = r(\cos\theta + i\sin\theta)) \\ &= \{0\} \cup \{w = \rho(\cos\phi + i\sin\phi) : \rho > 0, -\pi < \phi \le \pi\} \\ &\quad (\rho = r^3, \phi = 3\theta) \\ &= \mathbb{C}.\end{aligned}$$

Now, for each $w \in \mathbb{C}$ we wish to solve the equation
$$w = z^3 \qquad (1)$$
to obtain a unique solution z in A. If $w = 0$, then Equation (1) has the unique solution $z = 0$. On the other hand, if $w \ne 0$, then w can be written in the form
$$w = \rho(\cos\phi + i\sin\phi), \quad \text{where } \rho > 0 \text{ and } -\pi < \phi \le \pi,$$
and Equation (1) then has exactly three solutions:
$$z_0 = \rho^{1/3}\left(\cos\tfrac{1}{3}\phi + i\sin\tfrac{1}{3}\phi\right),$$
$$z_1 = \rho^{1/3}\left(\cos\left(\tfrac{1}{3}\phi + \tfrac{2\pi}{3}\right) + i\sin\left(\tfrac{1}{3}\phi + \tfrac{2\pi}{3}\right)\right)$$
and
$$z_2 = \rho^{1/3}\left(\cos\left(\tfrac{1}{3}\phi + \tfrac{4\pi}{3}\right) + i\sin\left(\tfrac{1}{3}\phi + \tfrac{4\pi}{3}\right)\right),$$
by Theorem 3.1 of *Unit A1*. Clearly $z_0 \in A$, since $-\pi/3 < \tfrac{1}{3}\phi \le \pi/3$, whereas z_1 and z_2 are not in A.
Thus f is a one-one function with image $\mathbb{C}$. Hence f has an inverse function f^{-1} with domain $\mathbb{C}$ and rule
$$f^{-1}(w) = \begin{cases} \rho^{1/3}\left(\cos\tfrac{1}{3}\phi + i\sin\tfrac{1}{3}\phi\right), & w = \rho(\cos\phi + i\sin\phi), \\ & \rho > 0, -\pi < \phi \le \pi, \\ 0, & w = 0.\end{cases}$$
(Since, for $w \ne 0$, $-\pi < \phi \le \pi$, $\phi = \operatorname{Arg} w$ and so z_0 is the principal cube root of w: $w^{1/3}$. Also $0^{1/3} = 0$, by definition, and so
$$f^{-1}(w) = w^{1/3} \quad (w \in \mathbb{C}).)$$

Section 2

2.1 (a) Since $f(z) = \bar{z} = x - iy$,
$$\operatorname{Re} f : z \longmapsto x \quad (z \in \mathbb{C})$$
and
$$\operatorname{Im} f : z \longmapsto -y \quad (z \in \mathbb{C}).$$
(b) Since $f(z) = iz = -y + ix$,
$$\operatorname{Re} f : z \longmapsto -y \quad (z \in \mathbb{C})$$
and
$$\operatorname{Im} f : z \longmapsto x \quad (z \in \mathbb{C}).$$
(c) Since $f(z) = z^3 = (x + iy)^3$
$$= x^3 - 3xy^2 + i(3x^2y - y^3),$$
$$\operatorname{Re} f : z \longmapsto x^3 - 3xy^2 \quad (z \in \mathbb{C})$$
and
$$\operatorname{Im} f : z \longmapsto 3x^2y - y^3 \quad (z \in \mathbb{C}).$$
(d) Since $f(z) = |z| = \sqrt{x^2 + y^2}$,
$$\operatorname{Re} f : z \longmapsto \sqrt{x^2 + y^2} \quad (z \in \mathbb{C})$$
and
$$\operatorname{Im} f : z \longmapsto 0 \quad (z \in \mathbb{C}).$$

2.2 (a) $s = \operatorname{Re}(\bar{z} + 1) = x + 1$

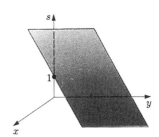

(b) $s = \operatorname{Im}(|z| + i) = 1$

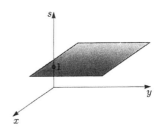

2.3 (a) $\gamma(t) = 1 - it \quad (t \in \mathbb{R})$

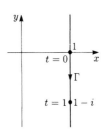

(b) $\gamma(t) = i + (1-i)t \quad (t \in [0,1])$

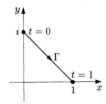

(c) $\gamma(t) = \cos t - i \sin t$
$= \cos(-t) + i \sin(-t) \quad (t \in [0, 2\pi])$

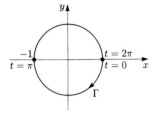

2.4 (a) $\gamma(t) = (1-t)(1+i) + ti$
$= 1 - t + i;$
so the parametric equations are
$$x = 1-t, \quad y = 1 \quad (t \in \mathbb{R}).$$
The path Γ is the line $y = 1$.

(b) From the table of standard parametrizations,
$$\gamma(t) = 2\cos t + 3i \sin t \quad (t \in [0, 2\pi])$$
is a parametrization of the ellipse
$$\frac{x^2}{4} + \frac{y^2}{9} = 1.$$

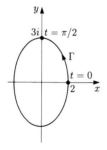

(c) Since $\gamma(t) = 1 + 2\cos t - (1 - 2\sin t)i$, the parametric equations are
$$x = 1 + 2\cos t, \quad y = -1 + 2\sin t \quad (t \in [0, 2\pi]).$$
Hence
$$(x-1)^2 = 4\cos^2 t \quad \text{and} \quad (y+1)^2 = 4\sin^2 t$$
and so
$$(x-1)^2 + (y+1)^2 = 4.$$
This is the equation of the circle with radius 2 and centre $1 - i$.

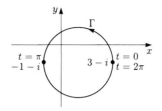

2.5 (a) $\gamma(t) = 1 - i + 3(\cos t + i \sin t) \quad (t \in [0, 2\pi])$.

(b) The equation $2x^2 + 3y^2 = 6$ is equivalent to
$$\frac{x^2}{3} + \frac{y^2}{2} = 1,$$
for which the standard parametrization is
$$\gamma(t) = \sqrt{3}\cos t + i\sqrt{2}\sin t \quad (t \in [0, 2\pi]).$$

(c) The equation $8y^2 = x$ is equivalent to
$$y^2 = \tfrac{1}{8}x,$$
for which the standard parametrization is
$$\gamma(t) = \tfrac{1}{32}t^2 + \tfrac{1}{16}it \quad (t \in \mathbb{R}).$$

2.6 $\gamma(t) = \tfrac{1}{2}(\cos t + i\sin t) - \tfrac{1}{4}(\cos 2t + i\sin 2t)$
$= \tfrac{1}{2}\cos t - \tfrac{1}{4}\cos 2t$
$\quad + i\left(\tfrac{1}{2}\sin t - \tfrac{1}{4}\sin 2t\right) \quad (t \in [-\pi, \pi]).$

Hence the table of values is as follows.

t	0	$\pm\tfrac{1}{4}\pi$	$\pm\tfrac{1}{2}\pi$	$\pm\tfrac{3}{4}\pi$	$\pm\pi$
x	0.25	0.35	0.25	-0.35	-0.75
y	0	± 0.10	± 0.5	± 0.60	0

Plotting these points, we obtain the following rough sketch of the path (which is a cardioid).

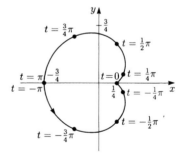

Since $x = \tfrac{1}{2}\cos t - \tfrac{1}{4}\cos 2t$ and $y = \tfrac{1}{2}\sin t - \tfrac{1}{4}\sin 2t$,
$$x^2 + y^2 = \left(\tfrac{1}{4}\cos^2 t - \tfrac{1}{4}\cos t \cos 2t + \tfrac{1}{16}\cos^2 2t\right)$$
$$+ \left(\tfrac{1}{4}\sin^2 t - \tfrac{1}{4}\sin t \sin 2t + \tfrac{1}{16}\sin^2 2t\right)$$
$$= \tfrac{1}{4}(\cos^2 t + \sin^2 t) - \tfrac{1}{4}(\cos t \cos 2t + \sin t \sin 2t)$$
$$+ \tfrac{1}{16}(\cos^2 2t + \sin^2 2t)$$
$$= \tfrac{1}{4} - \tfrac{1}{4}\cos(2t - t) + \tfrac{1}{16}$$
$$= \tfrac{5}{16} - \tfrac{1}{4}\cos t$$
$$= \tfrac{1}{16}(5 - 4\cos t).$$

Hence
$$4(x^2 + y^2)^2 - \tfrac{3}{2}(x^2 + y^2) + \tfrac{1}{2}x$$
$$= \tfrac{1}{64}(5 - 4\cos t)^2 - \tfrac{3}{32}(5 - 4\cos t)$$
$$+ \tfrac{1}{8}(2\cos t - \cos 2t)$$
$$= \left(\tfrac{25}{64} - \tfrac{5}{8}\cos t + \tfrac{1}{4}\cos^2 t\right) - \left(\tfrac{15}{32} - \tfrac{3}{8}\cos t\right)$$
$$+ \left(\tfrac{1}{4}\cos t - \tfrac{1}{8}(2\cos^2 t - 1)\right)$$
$$= \tfrac{3}{64}, \quad \text{as required.}$$

2.7 The principal square root function $f(z) = \sqrt{z}$ maps $z = r(\cos\theta + i\sin\theta)$, where $\theta = \operatorname{Arg} z$, to $w = r^{1/2}\left(\cos\tfrac{1}{2}\theta + i\sin\tfrac{1}{2}\theta\right)$.

(a) If z is on the negative x-axis, then $\theta = \pi$, and so its image w is such that $\operatorname{Arg} w = \pi/2$. Thus, the negative x-axis maps to the positive v-axis in the w-plane.

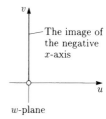

(b) The circle $|z| = 1$ comprises two disjoint parts:
U, the upper semi-circle, consisting of all points z such that $|z| = 1$ and $\operatorname{Arg} z \in [0, \pi]$;
L, the lower semi-circle, consisting of all points z such that $|z| = 1$ and $\operatorname{Arg} z \in]-\pi, 0[$.

Thus $f(U)$ is the quarter-circle in the w-plane given by $|w| = 1$, $\operatorname{Arg} w \in [0, \pi/2]$, and $f(L)$ is the quarter-circle in the w-plane given by $|w| = 1$, $\operatorname{Arg} w \in]-\pi/2, 0[$. Hence the image of the circle $|z| = 1$ is the semi-circle shown below.

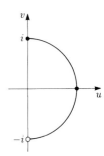

Section 3

3.1 In these solutions, the points labelled in the w-plane indicate the location of images of given grid lines. For example, in the first figure, the image of the line $y = 1$ is the line $v = 2$.

(a) The function $f(z) = z + i$ translates the point z one unit in the y direction. The images of the Cartesian grid and the polar grid are shown below.

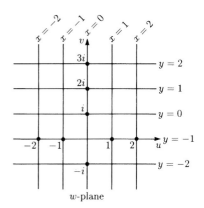

Image of Cartesian grid

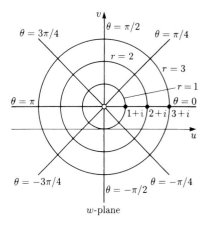

Image of polar grid

(b) The function $f(z) = 2z$ doubles the modulus of the point z, but leaves its argument unchanged. The images of the Cartesian grid and polar grid are shown below.

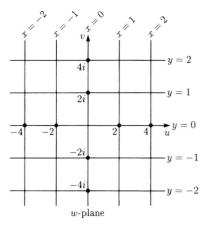

Image of Cartesian grid

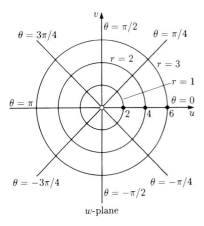

Image of polar grid

(c) The function $f(z) = 2 - iz$ rotates the point z about the origin through $\pi/2$ clockwise and then translates it 2 units to the right. Thus the images of the Cartesian grid and the polar grid are as follows.

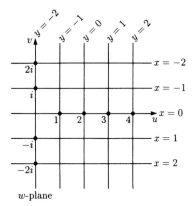

Image of Cartesian grid

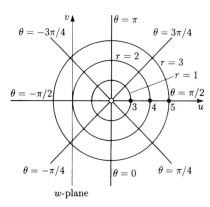

Image of polar grid

Alternatively, for the Cartesian grid we may use the parametric approach, as follows.

The image of z is
$$w = f(z) = 2 - iz$$
$$= 2 + y - ix,$$
so that
$$u = 2 + y, \quad v = -x. \quad (1)$$
The line $x = a$ has parametric equations
$$x = a, \quad y = t \quad (t \in \mathbb{R});$$
substituting these in Equations (1) gives the parametric equations of the image
$$u = 2 + t, \quad v = -a \quad (t \in \mathbb{R}),$$
which is the line $v = -a$.
Similarly, the line $y = b$ has parametric equations
$$x = t, \quad y = b \quad (t \in \mathbb{R});$$
substituting these in Equations (1) gives the parametric equations of the image
$$u = 2 + b, \quad v = t \quad (t \in \mathbb{R}),$$
which is the line $u = 2 + b$.

(d) Since $f(z) = iz^2 = i \times z^2$ and multiplication by i corresponds to rotation about the origin through $\pi/2$, the images are found by rotating those in Frames 7 and 8 through $\pi/2$.

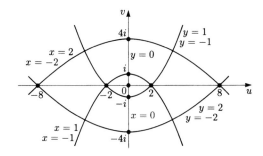

Image of Cartesian grid

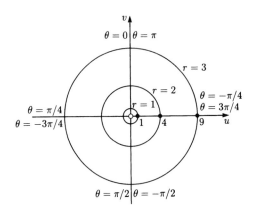

Image of polar grid

Section 4

4.1 (a) $e^{3\pi i} = e^0(\cos 3\pi + i \sin 3\pi)$
$$= -1$$
(b) $ee^{\pi i/2} = e(\cos \pi/2 + i \sin \pi/2)$
$$= ei$$
(c) $e^{2\pi i/3} = e^0(\cos 2\pi/3 + i \sin 2\pi/3)$
$$= -\frac{1}{2} + \frac{\sqrt{3}}{2}i$$
(d) $e^{-3\pi i/2} = e^0(\cos(-3\pi/2) + i\sin(-3\pi/2))$
$$= i$$
(e) $e^{2+\pi i} = e^2(\cos \pi + i \sin \pi)$
$$= -e^2$$
(f) $e^{3+\pi i/2} = e^3(\cos \pi/2 + i \sin \pi/2)$
$$= e^3 i$$
(g) $e^{(\pi i/6)-1} = e^{-1}(\cos \pi/6 + i \sin \pi/6)$
$$= \frac{\sqrt{3}}{2e} + \frac{1}{2e}i$$
(h) $e^{(\cos \theta + i \sin \theta)} = e^{\cos \theta}(\cos(\sin \theta) + i \sin(\sin \theta))$

4.2 (a) (i) $\left|\frac{1}{\sqrt{2}} - \frac{i}{\sqrt{2}}\right| = 1$ and
$\text{Arg}\left(\frac{1}{\sqrt{2}} - \frac{i}{\sqrt{2}}\right) = -\frac{\pi}{4}$; hence
$$\frac{1}{\sqrt{2}} - \frac{i}{\sqrt{2}} = 1e^{-\pi i/4} = e^{-\pi i/4}.$$

(ii) $|-(1+i)| = \sqrt{2}$ and $\text{Arg}(-(1+i)) = -3\pi/4$; hence
$-(1+i) = \sqrt{2}e^{-3\pi i/4}.$

(iii) $|1 + \sqrt{3}i| = 2$ and $\text{Arg}(1 + \sqrt{3}i) = \pi/3$; hence
$1 + \sqrt{3}i = 2e^{\pi i/3}.$

(b) (i) $\left(\frac{1}{\sqrt{2}} - \frac{i}{\sqrt{2}}\right)^3 = (e^{-\pi i/4})^3$
$$= e^{-3\pi i/4}$$
$$= -\frac{1}{\sqrt{2}} - \frac{i}{\sqrt{2}}$$

(ii) $(1 + \sqrt{3}i)^{-7} = (2e^{\pi i/3})^{-7}$
$$= 2^{-7}e^{-7\pi i/3}$$
$$= 2^{-7}e^{-\pi i/3} \quad (\text{exp has period } 2\pi i)$$
$$= 2^{-7}\left(\frac{1}{2} - \frac{\sqrt{3}}{2}i\right)$$
$$= 2^{-8}(1 - \sqrt{3}i).$$

4.3 (a) $\sin(\pi + 2i) = \frac{1}{2i}(e^{i(\pi+2i)} - e^{-i(\pi+2i)})$
$$= \frac{1}{2i}(e^{-2+i\pi} - e^{2-i\pi})$$
$$= \frac{1}{2i}(e^{-2}e^{i\pi} - e^2 e^{-i\pi})$$
$$= \frac{1}{2i}(-e^{-2} + e^2) = -\left(\frac{e^2 - e^{-2}}{2}\right)i.$$

Alternatively, we have
$\sin(\pi + 2i) = \sin\pi\cos(2i) + \cos\pi\sin(2i)$
$\quad\quad\quad\quad\quad\quad$ (Theorem 4.3(a))
$= 0 + (-1) \times (i\sinh 2) \quad$ (Theorem 4.4)
$= -\left(\dfrac{e^2 - e^{-2}}{2}\right)i.$

(b) $\cos(\pi/2 - i) = \frac{1}{2}(e^{i(\pi/2-i)} + e^{-i(\pi/2-i)})$
$= \frac{1}{2}(e^{1+i\pi/2} + e^{-1-i\pi/2})$
$= \frac{1}{2}(ee^{i\pi/2} + e^{-1}e^{-i\pi/2})$
$= \frac{1}{2}(ei - e^{-1}i)$
$= \left(\dfrac{e - e^{-1}}{2}\right)i \ (= i\sinh 1).$

Alternatively, we have
$\cos(\pi/2 - i) = \cos(\pi/2)\cos(-i) - \sin(\pi/2)\sin(-i)$
$\quad\quad\quad\quad\quad$ (Theorem 4.3(a))
$= 0 - 1 \times (i\sinh(-1)) \quad$ (Theorem 4.4)
$= i\sinh 1 \quad$ (Theorem 4.5(c))
$= \left(\dfrac{e - e^{-1}}{2}\right)i.$

(c) $\tan i = \dfrac{\sin i}{\cos i}$
$= \left(\dfrac{1}{2i}(e^{i^2} - e^{-i^2})\right) \Big/ \left(\dfrac{1}{2}(e^{i^2} + e^{-i^2})\right)$
$= \dfrac{e^{-1} - e}{(e^{-1} + e)i}$
$= \left(\dfrac{e - e^{-1}}{e + e^{-1}}\right)i \ (= i\tanh 1).$

4.4 (a) $\overline{e^z} = \overline{e^{x+iy}} \quad (z = x + iy)$
$= \overline{e^x e^{iy}}$
$= \overline{e^x}\,\overline{e^{iy}} \quad (e^x \text{ is real})$
$= e^x \overline{(\cos y + i\sin y)}$
$= e^x(\cos y - i\sin y)$
$= e^x(\cos(-y) + i\sin(-y))$
$= e^x e^{-iy}$
$= e^{x-iy} = e^{\bar z}$

(b) $\sin 2z = \dfrac{1}{2i}(e^{2iz} - e^{-2iz})$
$= \dfrac{1}{2i}((e^{iz})^2 - (e^{-iz})^2)$
$= \dfrac{1}{2i}(e^{iz} + e^{-iz})(e^{iz} - e^{-iz})$
$= \dfrac{1}{2i}(2\cos z)(2i\sin z)$
$= 2\sin z\cos z$

(c) $\overline{\sin z} = \overline{\dfrac{1}{2i}(e^{iz} - e^{-iz})}$
$= \dfrac{1}{\overline{2i}}(\overline{e^{iz}} - \overline{e^{-iz}}) \quad$ (*Unit A1*, Theorem 1.1(b))
$= \dfrac{1}{2(-i)}(e^{\overline{iz}} - e^{\overline{-iz}}) \quad$ (part (a))
$= -\dfrac{1}{2i}(e^{-i\bar z} - e^{i\bar z})$
$= \dfrac{1}{2i}(e^{i\bar z} - e^{-i\bar z})$
$= \sin\bar z$

(d) $\cosh(z_1 + z_2) = \cos(i(z_1 + z_2)) \quad$ (Theorem 4.4)
$= \cos(iz_1 + iz_2)$
$= \cos(iz_1)\cos(iz_2) - \sin(iz_1)\sin(iz_2)$
$\quad$ (Theorem 4.3(a))
$= \cosh z_1 \cosh z_2 - (i\sinh z_1)(i\sinh z_2)$
$\quad$ (Theorem 4.4)
$= \cosh z_1 \cosh z_2 + \sinh z_1 \sinh z_2$

(e) $\cosh^2 z - \sinh^2 z = \cos^2(iz) - i^2\sin^2(iz) \quad$ (Theorem 4.4)
$= \cos^2(iz) + \sin^2(iz)$
$= 1 \quad$ (Theorem 4.3(b))

4.5 By Example 4.4(a),
$\sin z = \sin x\cosh y + i\cos x\sinh y;$
hence if $w = u + iv = \sin z$, then
$$u = \sin x\cosh y \quad \text{and} \quad v = \cos x\sinh y. \tag{1}$$
The line $x = a$, $y \geq 0$ (where $0 \leq a \leq \pi/2$) has parametric equations
$$x = a, \quad y = t \quad (t \in [0, \infty[);$$
substituting these in Equations (1), we obtain the parametric equations of the image
$$u = \sin a\cosh t, \quad v = \cos a\sinh t \quad (t \in [0, \infty[). \tag{2}$$

Using the table of standard parametrizations, we identify the image of the line $x = a$ as the hyperbola (first quadrant portion)
$$\frac{u^2}{\sin^2 a} - \frac{y^2}{\cos^2 a} = 1,$$
provided that $\sin a \neq 0$ and $\cos a \neq 0$ (true for $0 < a < \pi/2$).

If $a = 0$, then from Equations (2), the image is the non-negative v-axis.

If $a = \pi/2$, then from Equations (2), the image is the positive u-axis, excluding the interval $]0, 1[$.

The line $y = b$, $0 \leq x \leq \pi/2$ (where $b \geq 0$) has parametric equations
$$x = t, \quad y = b \quad (t \in [0, \pi/2]);$$
substituting these in Equations (1), we obtain the parametric equations of the image
$$u = \sin t \cosh b, \quad v = \cos t \sinh b \quad (t \in [0, \pi/2]). \quad (3)$$
Using the table of standard parametrizations, we identify the image of the line $y = b$ as the ellipse (first quadrant portion)
$$\frac{u^2}{\cosh^2 b} + \frac{v^2}{\sinh^2 b} = 1,$$
provided that $\sinh b \neq 0$ (true for $b > 0$).

If $b = 0$, then from Equations (3), the image is the interval $[0, 1]$ of the u-axis.

Using the above information about the images, we obtain the image of the given grid as follows. (The images of other horizontal and vertical lines $x = a$, $y = b$ are easy to visualize.)

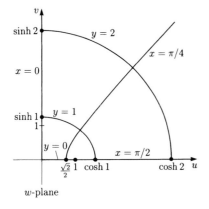

Section 5

5.1 (a) $\text{Log}(-2) = \log_e |-2| + i\,\text{Arg}(-2)$
$= \log_e 2 + i\pi$

(b) $\text{Log}(i^3) = \text{Log}(-i)$
$= \log_e |-i| + i\,\text{Arg}(-i)$
$= -i\pi/2$

(c) $\text{Log}(1+i) = \log_e |1+i| + i\,\text{Arg}(1+i)$
$= \log_e \sqrt{2} + i\pi/4$

(d) $\text{Log}\sqrt{3} = \log_e \sqrt{3} + i\,\text{Arg}\sqrt{3}$
$= \log_e \sqrt{3}$

(e) $\text{Log}(i - \sqrt{3}) = \log_e |i - \sqrt{3}| + i\,\text{Arg}(i - \sqrt{3})$
$= \log_e 2 + i5\pi/6$

(f) $\text{Log}\left(\dfrac{1-i}{\sqrt{2}}\right) = \log_e \left|\dfrac{1-i}{\sqrt{2}}\right| + i\,\text{Arg}\left(\dfrac{1-i}{\sqrt{2}}\right)$
$= \log_e 1 + i(-\pi/4)$
$= -i\pi/4$

5.2 (a) $i^{-i} = \exp(-i\,\text{Log}\,i)$
$= \exp(-i(0 + i\pi/2))$
$= e^{\pi/2}$

(b) $(-i)^i = \exp(i\,\text{Log}(-i))$
$= \exp(i(0 - i\pi/2))$
$= e^{\pi/2}$

(c) $(1-i)^i = \exp(i\,\text{Log}(1-i))$
$= \exp(i(\log_e |1-i| + i\,\text{Arg}(1-i)))$
$= \exp(i\log_e(\sqrt{2}) + \pi/4)$
$= e^{\pi/4}(\cos(\log_e \sqrt{2}) + i\sin(\log_e \sqrt{2}))$

(d) $(1+i)^{1+i} = \exp((1+i)\text{Log}(1+i))$
$= \exp((1+i)(\log_e |1+i| + i\,\text{Arg}(1+i)))$
$= \exp((1+i)(\log_e \sqrt{2} + i\pi/4))$
$= \exp(\log_e(\sqrt{2}) - \pi/4 + (\log_e(\sqrt{2}) + \pi/4)i)$
$= e^{\log_e(\sqrt{2}) - \pi/4}(\cos(\log_e(\sqrt{2}) + \pi/4)$
$\quad + i\sin(\log_e(\sqrt{2}) + \pi/4))$
$= \sqrt{2}e^{-\pi/4}(\cos(\log_e(\sqrt{2}) + \pi/4)$
$\quad + i\sin(\log_e(\sqrt{2}) + \pi/4))$

(e) $(-1)^i = \exp(i\,\text{Log}(-1))$
$= \exp(i(0 + i\pi))$
$= e^{-\pi}$

(f) $(-1)^2 = 1 \quad (-1 \text{ is real})$

(Applying the definition of principal power, we obtain
$(-1)^2 = \exp(2\,\text{Log}(-1))$
$= \exp(2(0 + i\pi))$
$= e^{2\pi i} = 1.$)